FUNCTIONAL AND CAUSAL ASPECTS OF NEST DISTRIBUTION IN COLONIES OF THE SANDWICH TERN

(*STERNA S. SANDVICENSIS* LATH.)

BEHAVIOUR

AN INTERNATIONAL JOURNAL OF COMPARATIVE ETHOLOGY

SUPPLEMENT XX

J. VEEN

FUNCTIONAL AND CAUSAL ASPECTS OF NEST DISTRIBUTION IN COLONIES OF THE SANDWICH TERN

(*STERNA S. SANDVICENCIS* LATH.)

LEIDEN
E. J. BRILL
1977

FUNCTIONAL AND CAUSAL ASPECTS OF NEST DISTRIBUTION IN COLONIES OF THE SANDWICH TERN

(*STERNA S. SANDVICENSIS* LATH.)

BY

J. VEEN

Zoological Laboratory, University of Groningen

With 51 figures and 25 plates

LEIDEN
E. J. BRILL
1977

ISBN 90 04 04906 1

PRINTED IN THE NETHERLANDS

CONTENTS

Part I. GENERAL ASPECTS

Part II. ASPECTS OF FUNCTION

PART I

GENERAL ASPECTS

1. INTRODUCTION

1.1. THE PROBLEM

Among the many characteristics of a species of bird is its breeding distribution. Some species tend to nest evenly distributed over the area occupied in the nesting period. Others are concentrated within a relatively small part of the breeding area; they nest in colonies.

During the last decades an increasing number of workers has considered the way in which nest distribution might contribute to survival. Special attention has been paid to colonial nesting. The results obtained have shown that colonial nesting involves a complex relationship of aspects of behaviour and ecology; the ways in which it might influence reproductive success are manifold.

Colonial nesting might contribute to success through facilitation of pair formation (cf. sea-birds in general, LACK, 1968), synchronization of chick hatching to favourable food conditions (Atlantic gannet, NELSON, 1966, 1970; tropical terns, ASHMOLE, 1962, 1963b), exchange of information about food (cf. blue heron, KREBS, 1974; CROOK, 1965; LACK, 1968; WARD & ZAHAVI, 1973), communal defense of the breeding area against predators (cf. blackheaded gull, KRUUK, 1964). However, colonial nesting might be detrimental as a consequence of food- and space competition, spread of diseases and parasites and conspicuousness to predators.

The present study is an attempt to investigate whether in the sandwich tern colonial nesting enhances reproductive success, and if so how this comes about.

Considering the various ways in which colonial nesting might influence success—in fact all aspects listed above might be relevant in the case of the sandwich tern—a study of function would ideally require comprehensive study of all environmental factors. Although the relation between nest distribution and several environmental factors will be touched upon in this study, my concern has been primarily with predation. The choice of predation as the main subject of study has been inspired by the work done by TINBERGEN and co-workers studying nest distribution in gulls.

Several years ago N. TINBERGEN had already noted that the spatial pattern of nest distribution in a gull colony is non-random. On the one hand there was a clear tendency to cluster, but on the other short inter-nest distances were underrepresentated. Using an experimental approach (TINBERGEN *et al.*, 1962, 1967; CROZE, 1970) it was shown that nest spacing assists concealment of the nests, which in turn lowers the risk that eggs and chicks are taken by predators. KRUUK (1964) studying predators and anti-predator behaviour of the black-headed gull showed that clustering has the advantage that the predator can be simultaneously attacked by a large number of breeding birds. In the black-headed gull such mass attacks turned out to be a very effective defense of the brood.

So, according to TINBERGEN and co-workers, the spatial pattern of nest-distribution in a black-headed gull colony should be regarded as a consequence of two opposing tendencies, clustering and spacing out, which both have an anti-predator function. The observed pattern would be the best compromise between both tendencies, considering the circumstances in which the species lives. Whether or not the observed pattern is indeed the best compromise has been tested by PATTERSON (1965). It followed from his study that colony nesting black-headed gulls raised more chicks than solitary nesting individuals. There also was a strong correlation between synchronization of egg-laying and breeding success, but no correlation was found between breeding success and nest density.

Similarities in the behaviour and nest distribution pattern of black-headed gulls on the one hand and great and lesser black-headed gulls, herring gulls, common gulls, arctic terns, common terns, roseate terns and little terns on the other, suggest that the majority of tern and gull species of the northern hemisphere defend their brood against predators in a basically similar way. However, there are some exceptions: In the kittiwake nests are crowded side by side. Camouflage is poorly developed and aggressive behaviour directed towards predators is nearly absent. The deviant behaviour and nest distribution of the kittiwake have been explained by the absence of predators in the nesting area (CULLEN, 1957).

The sandwich tern is another species with a nest distribution pattern which is markedly different from that of most other tern and gull species. Sandwich terns nest at pecking distance of each other (Plate 6). Although eggs and chicks have a cryptic colouration, one cannot escape the impression that the colony area is rather conspicuous. It has also been reported that the sandwich tern is not very inclined to attack predators.

In contrast to the situation in the kittiwake this cannot be explained by the absence of predators, for a variety of predator species have been recorded from sandwich tern colonies (CULLEN, 1960a; DIRCKSEN, 1932; KRUUK, 1964; LIND, 1963; ROOTH, 1958, 1965).

In case of the sandwich tern there is another point of interest which might be related to predation. Several authors have drawn attention to the fact that sandwich terns often nest in association with common terns, arctic terns and black-headed gulls (Plate 5) (CULLEN, 1960a; LIND, 1963; ROOTH, 1965). It has been suggested that the sandwich terns might gain protection against predators through the aggressiveness of these neighbours. However, it has also been noted (VAN DEN ASSEM, 1954b; DE SMIDT & BROEKHUIZEN, 1964; ROOTH, 1965) that the black-headed gull takes a heavy toll from the terns by stealing eggs, chicks and fish. Thus, as far as the black-headed gull is concerned sandwich terns seem to nest with a species which might provide some protection although being a predator itself.

Summarizing the above information we may conclude that the nest distribution pattern of the sandwich tern is different from other Larids with respect to a number of aspects which in the latter may be interpreted as anti-predator behaviour. However, sandwich terns often nest in association with arctic terns, common terns and black-headed gulls. As a consequence, they will meet the same predators. This raises the question how sandwich terns defend themselves against predators, and why they nest at densities far greater than is typical for most other larids. The present study is primarily concerned with the first question. The second question is only briefly touched upon. Finally, in addition to the above questions of function, some attention has been paid to the causation of the nest distribution pattern observed.

1.2. ORGANIZATION OF THE PAPER

Chapter I.2. gives general information on a number of aspects of behaviour and ecology of the sandwich tern, which might be important for this study. It deals with: breeding chronology, colony structure, site selection and food.

Part II of this study deals with functional aspects of nest distribution.

In Chapter II.1. an attempt is made to answer the question 'whether or not nest distribution or aspects related to it, contribute to reproductive success'. It consists of an analysis of aspects of aggregation and synchronization, clutch size and age in relation to reproductive success. Besides, it touches upon a number of mortality factors to which these aspects might be adapted.

Chapter II.2. gives a survey of the occurrence of predators, their hunting behaviour and the actual predation on the sandwhich tern. In two seasons the influence of predation by black-headed gulls, herring gulls and great black-backed gulls on the breeding success of the sandwich tern is determined.

Chapter II.3. deals with how sandwich terns defend their brood against predators. A survey of the anti-predator reactions of sandwich tern, common tern, arctic tern and black-headed gull is given. The effects of these reactions upon the predators, and in case of common tern, arctic tern and black-headed gull its implications for the sandwich tern, are discussed.

In Chapter II.4. we will discuss (a) The various aspects of nest distribution in relation to predation, (b) the balance of advantage and disadvantage of associative nesting and (c) the various environmental pressures to which the sandwich tern nest distribution pattern might be adapted.

Part III deals with aspects of causation of nest distribution. An attempt will be made to investigate (a) what behaviour patterns lead to the specific nest distribution pattern within a colony and (b) what behaviour patterns lead to associative nesting.

1.3. THE STUDY AREA

The small isle of Griend is situated in the Dutch Waddensea, halfway between the Frisian coast and the island Terschelling (see Fig. 1). Griend is the highest part of the 'Grienderwaard', an extensive area of tidal flats, surrounded by deep channels in the North and West. The isle itself (Fig. 2, Plate 1 and 2), defined as the area above the line of mean high tide, measured 20 to 15 ha between 1965 and 1972. It consists of a relatively low and marshy inner part, densely vegetated with halophytous plants (mainly *Limonium vulgare, Halimione portulacoides, Artemisia maritima* and *Atriplex littoralis*), surrounded by a somewhat higher sandy ridge, which is only sparsely covered (mainly by *Elytrigia pungens, E. junceum* and *Elymus arenaria*).

Sometimes large areas or even the whole island are flooded, when long lasting strong winds from the North and West cause a rise of the water level. This occasionally happens in summer, but is quite normal in winter. The isle as a whole is highly subject to changes caused by such tides.

Sandwich terns, common terns and arctic terns have been reported to nest on the island from the beginning of this century (BROUWER *et al.*,

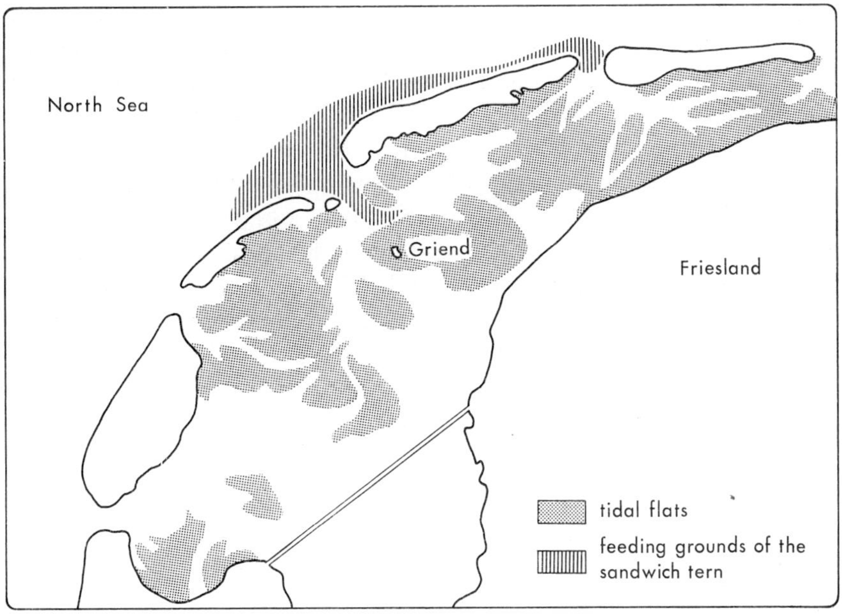

Fig. 1. Map of the western part of the Waddensea.

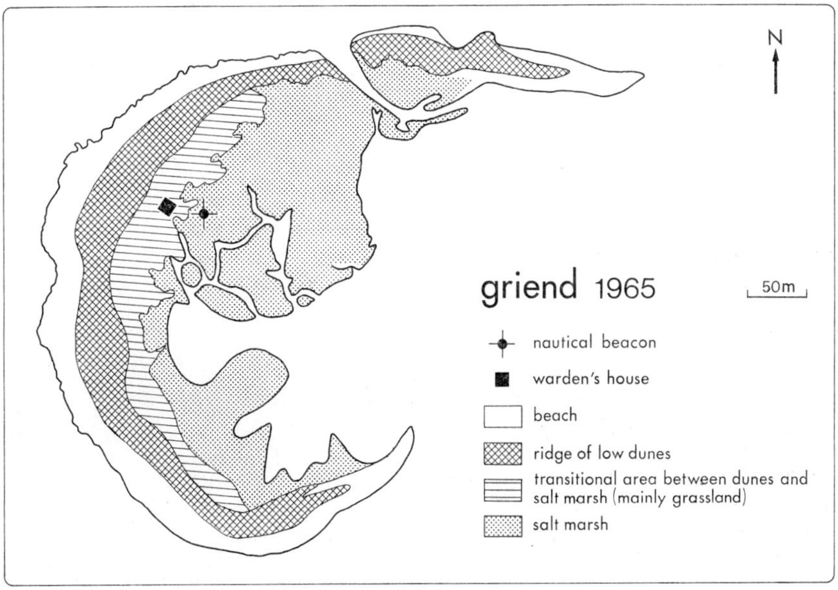

Fig. 2. Topography of the island of Griend.

1950). Egg-collecting, however, probably prevented any successful breeding until the island became protected as a nature reserve in 1915. Thereupon numbers increased considerably with peaks of more than 20.000 pairs of sandwich terns (1932-1939), 15.000 pairs of common terns (1946, 1954) and 2000 pairs of arctic terns (1954). In 1964 and 1965 numbers dropped markedly as a result of pollution by organochlorine pesticides (KOEMAN *et al.*, 1967, 1968; VEEN, 1967). Ever since, the tern populations on the island have been relatively stable (see Table 1). Black-headed gulls first settled in 1921. To protect the terns,

TABLE 1

Numbers of breeding pairs of sandwich tern, common tern, arctic tern and black-headed gull on Griend

	Sandwich tern	Common tern	Arctic tern	Black-headed gull
1964	1500	1200	300	170
1965	800	1200	300	300
1966	1500	1200	300	700
1967	1000	1100	350	800
1968	1200	800	300	1200
1969	1600	1000	600	3000
1970	1700	800	600	4000
1971	2100	900	500	4000
1972	2100	800	500	4000
1973	2200	700	600	4000
1974	2700	1000	450	4000
1975	2250	1400	600	6000

eggs have been collected between 1928 and 1963 and their numbers never exceeded a few hundred pairs. Recently their numbers have increased enormously.

1.4. General study techniques

The study has been carried out entirely in the field. All observations have been made on the isle of Griend. Within a radius of 15 km of the island the feeding area of the terns has been visited by boat. Short distance observations have been made from a blind, either placed on the ground or on a platform varying in height between 1 and 2 meters. The magnification of the binoculars used varied between 7x and 40x.

The main part of the study consists of observations done under natural circumstances. Since under natural conditions it is sometimes prohibitively time-consuming to accumulate sufficient observations, a series of field experiments were included.

1.5. Acknowledgements.

I am greatly indebted to Prof. Dr. G. P. BAERENDS, who introduced me to the study of animal behaviour. Without his keen interest, encouragement and critical comments this study would not have been accomplished.

Many thanks are due to my wife, Mrs H. J. VEEN-DALLMEIJER, who assisted me in the field in four years of observation. Her assistence, patience and understanding during all stages of this study have been of great help.

I am indebted to Prof. Dr W. H. VAN DOBBEN, Dr R. H. DRENT and Dr H. KRUUK for advice during my work and for critical comments on the manuscript.

Thanks are also due to scientists and techniciens of the Zoology Laboratory of Groningen University; in particular to Mrs H. LOCHORN-HULSEBOS who typed the successive versions of the manuscript, Mrs G. BAEYENS who made the French summary, Mr D. VISSER who prepared the figures, Mr L. HOEKSTRA who prepared the tern and gull dummies and to Dr H. REDDINGIUS who helped with statistical problems.

It is almost impossible to mention all those who have helped with the work in the field; however, I would like to thank in particular Mr H. FEY, Mr G. JAMBROES and Mr J. VAN DE KAM.

Mr J. BEIJERSBERGEN, Mr K. BOOT, Mr S. BRAAKSMA, Mr T. LEBRET, Mr H. T. van der MEULEN and Mr J. ROOTH allowed me to quote unpublished data on the sandwich tern. The Koninklijk Nederlands Meteorologisch Instituut (K.N.M.I.) provided wind-speed data measured on the island. The late Mr E. KOSTER en Mr O. H. M. KOSTER maintained the boat connection between island and mainland. Mr H. P. GORTER and Ir P. J. van HERWERDEN of the Vereniging tot Behoud van Natuurmonumenten in Nederland kindly allowed me to work on Griend. I wish to thank them all very much.

The work was supported by a grant from the Nederlandse Organisatie voor Zuiver Wetenschappelijk Onderzoek (Z.W.O.) through the medium of the Stichting Wetenschappelijk Duinonderzoek (S.W.D.).

2. NATURAL HISTORY DATA FOR THE SANDWICH TERN AT GRIEND

2.1. BREEDING CHRONOLOGY

Arrival.

The sandwich terns of the NW-European population spend the northern winter along the West coast of Africa. In the Netherlands they return from the end of March until the beginning of June (Commissie voor de Ned. Avifauna, 1962).

Usually I could not reach the study area before the end of April. Then the terns were already present in large numbers. In 1968 and 1970, however, observations were started on the 5th and the 7th of April, which made it possible to study arrival and occupation of the breeding area from the very beginning. Figure 3 gives for both years the daily maximum number of birds present in the neighbourhood of the island. Only a few individuals were seen in the beginning of April. Numbers increased considerably in the middle of the month, thereupon followed by a more gradual increase.

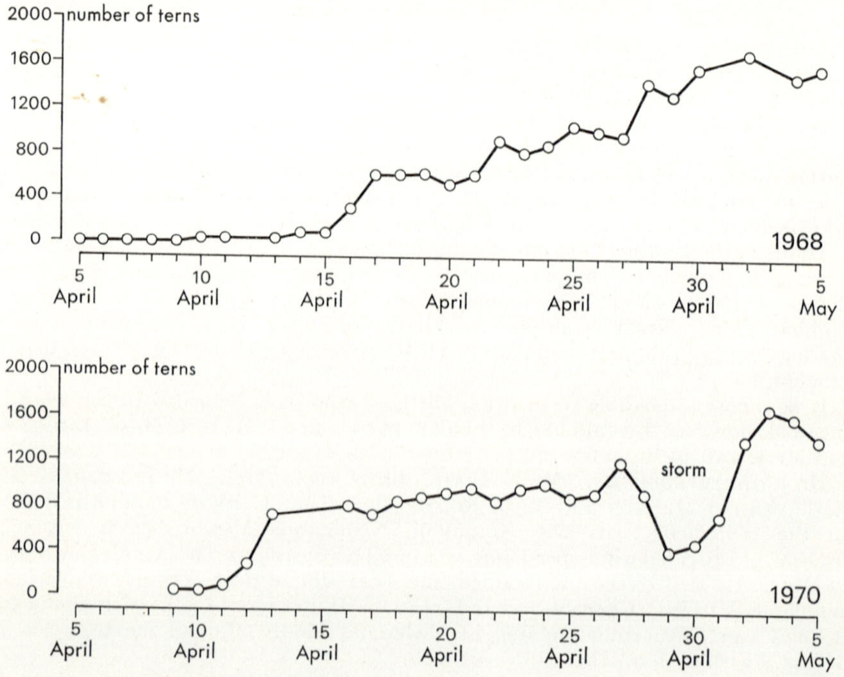

Fig. 3. The daily maximum number of sandwich terns present in the neighbourhood of the island in the pre-laying period in 1968 and 1970.

In all observation years the majority of birds was present on the island at the end of April. Usually relatively small numbers were still arriving in May and June. Late arrival of large numbers was only observed in May 1970 (40% of the breeding population) and in the beginning of June 1972 (30% of the breeding population).

Diurnal rhythm.

In the pre-laying period the number of terns on the island varied greatly at different times of the day, with peaks in the early morning and late evening (Fig. 4). In day-time most of the terns foraged at sea. In the evening, between 18.00 and 20.00 hours they concentrated near the island to spend the night in large flocks. Early in the morning, before sunrise, they disappeared again to their fishing grounds.

In the beginning of April the tidal flats near the island were used for sleeping. Later on the terns shifted to the dune-like ridge of the island which is the actual nesting area. In 1968 as well as in 1970 I saw a few birds landing on this ridge in the middle of the day, followed by a mass-

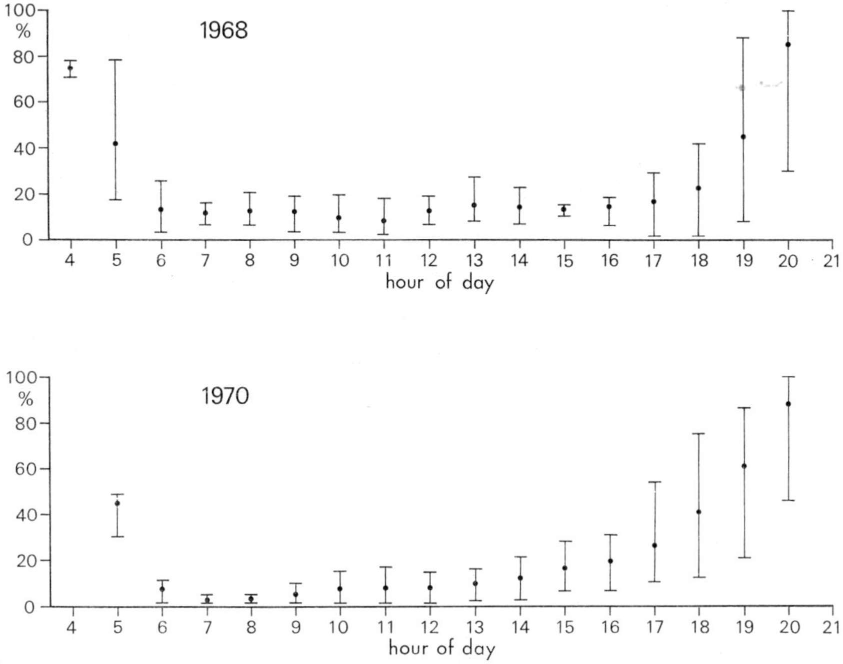

Fig. 4. The presence of sandwich terns on Griend at different times of the day in the pre-laying period, expressed as percentage of daily total, mean (black dots) and total range indicated.

occupation of the area for sleeping in the evening. In both years this happened three to four days after the sudden arrival of hundreds of birds in the middle of April (Fig. 3). Once having occupied the island for sleeping, up to a hundred terns might visit the area in day-time. However, until a few days before the start of egg-laying this happened very irregularly.

Courtship display and pair formation.

Courtship displays started as soon as the first birds had arrived in the breeding area. In the beginning of April aerial displays dominated, but in the second half of the month ground displays gradually came instead.

Display activity strongly depended on weather circumstances; warm and sunny weather always caused a considerable increase. We also noticed a difference in display activity at different times of the day. The relatively small number of birds visiting the island in day-time usually slept or preened, while only a small proportion was involved in courtship displays. However, when the terns started gathering on the island in the late afternoon, there usually was a marked increase of the display activity. In general, the terns all gathered in one or two groups of high density. Usually, the display activity remained high until shortly before sunset, after which there was a rapid decrease.

In the pre-laying period it was not possible to follow the terns individually. This made it difficult to establish whether or not at a certain moment they were mated. However, in the beginning of April 1968 and 1970 we saw a surprisingly high proportion of the terns flying around in couples. On 20 April 1968 more than 70% of 520 birds flew around in pairs. If we realise that the majority of these birds had arrived in the neighbourhood of the island only a few days before, it seems likely that most of them were already mated before they arrived in the breeding area.

Panic flights.

Among the sandwich terns present on the island in April and May social flights, (generally termed panics), were a common event. The performance of these flights can be described as follows: all birds of a group suddenly fly up and away towards the open flats or the sea. The members of a group often make synchronized swerving movements. The whole performance may last from seconds to a few minutes and the otherwise silent action usually ends in a noisy return at the original spot. Sometimes a panic which had started on one end of the island caused a chain of reactions as far as to the other end. Very often sandwich terns, common terns, arctic terns and black-headed gulls were involved in panics at the same time.

Panic flights were common in the beginning of the season, but they soon disappeared after the terns had started egg-laying (Fig. 5). On a few occasions panic flights were seen in day-time. Usually, however, they were recorded in the evening when the terns had concentrated on the island in large numbers. In general the number of panics reached a peak just before sunset.

Settlement, eggs and chicks.

From the beginning of May onwards the number of terns present on the island in day-time strongly increased, groups of terns suddenly appearing on places which had been previously occupied for sleeping. Within these groups displays were very prominent and for the first time indi-

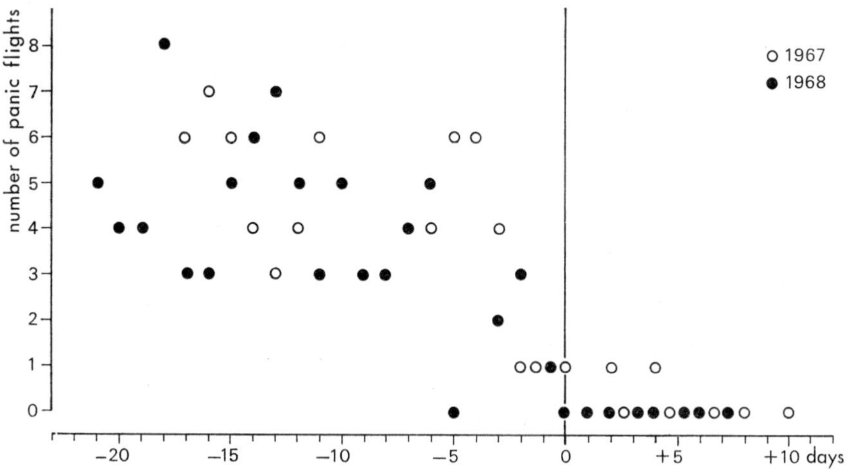

Fig. 5. The daily number of panic flights recorded between 18.00 and 20.00 hours in relation to the date on which 10% of the sandwich tern population had started egg-laying.

viduals defended a territory. Within a few days after appearance the majority of the terns of a group started egg-laying. In succession several groups might settle from the beginning of May till the end of July. Usually several groups settled next to each other.

Sandwich terns nest at pecking distance of each other. The nest consists of a simple scrape in the sand. One or two eggs are laid. Three-egg clutches have been recorded (less than one per cent), but they probably all originate from more than one female. In two-egg clutches the eggs are laid with an interval of 2 to 4 days. Incubation starts after the first egg has been laid and, as a consequence both eggs hatch asynchronously (interval 1 to 5 days, mean 2.45). Incubation lasts from 22 to 26 days. During the first days of incubation the female takes care of the brood. She is fed on the nest by the male. Later on both partners participate in incubation.

In contrast to other larid species (*e.g.* the black-headed gull) the sandwich tern defecates near its nest (Plate 7). At the end of the incubation period the colony is a white spotted area which often contrasts with the dark vegetation in the neighbourhood. Against the 'white painted' colony background the predominantly white egg has a cryptic colouration.

Within a week after hatching the chicks are led away by the parents to the more vegetated area which usually surrounds the colony (Plate 9).

As a consequence of synchronized egg-laying groups of breeding birds leave the nesting area more or less at the same time. Once having left the colony, the chicks may spread over a relatively large area.

During the first week after leaving the chicks are constantly guarded by one of both parents. Thereupon they are left alone more and more, while the parents make their fishing trips. Chicks are able to fly at the age of 4 weeks. When 6 weeks old they usually leave the island. On Griend they have all disappeared from the nesting area by about the middle of August. Till the end of September groups of birds, both adults and juveniles, may concentrate on the flats around the island to spend the night. However, ring recoveries have shown that these birds do not necessarily originate from the Griend breeding population.

2.2. COLONY STRUCTURE

In all years of observation the following units have been distinguished in the nest distribution pattern of the sandwich tern:

(1) groups (consisting of a varying number of nests),
(2) colonies (consisting of a varying number of groups),
(3) island population (consisting of a varying number of colonies).

It has been emphasized before that in the sandwich tern the occupation of the nesting territory is a remarkably quick procedure: groups of birds suddenly appear on the island in day-time; individual pairs immediately occupy a territory and within a few days after appearance the first egg is laid. In all years of observation 15 to 26 groups varying in size between 15 and 216 breeding pairs, settled on the island. The majority of these groups appeared with short intervals during the month of May, whereas the remainder settled in the course of June and July.

The temporal distribution of egg-laying of 11 groups which settled on the island in 1971 is given in Figure 6. It shows, that for each group egg-laying was markedly synchronized; in general 80-90% of the total number of pairs of a group started egg-laying within a 2-5 day period, followed by a relatively small number of pairs spread over a period of 1-3 weeks. The figure also shows the relatively small, but nevertheless distinct difference in the mean date of egg-laying between the various groups.

Groups appearing on the island with short intervals usually settled at short distances from each other. Although the nests of such groups might approach each other very closely, neighbouring groups never mingled and a group of new-comers never encircled another group but always settled

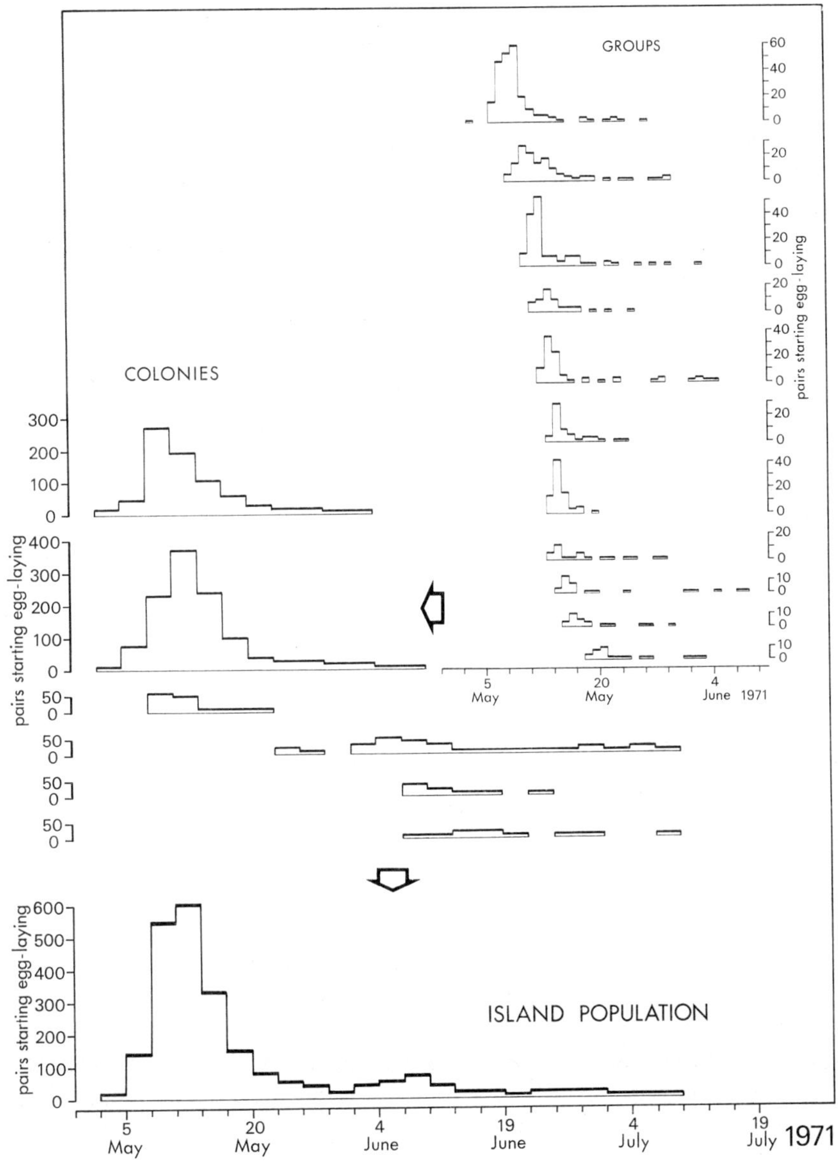

Fig. 6. Distribution of egg-laying of the island population in 1971 in total and specified for the six colonies of which this population consisted and for all groups of the second colony.

next to it. So, even when several groups concentrated within the same area, some spatial isolation was maintained. Groups which settled a week or more later than other groups, nearly always occupied an area which was at a distance of at least some tens of meters from the other groups already present.

In the course of the season a number of concentrations of breeding birds, thus appeared on the island, all spatially separated from each other. Such concentrations, irrespective of whether they were composed of only one or several groups, have been termed colonies. Figure 7

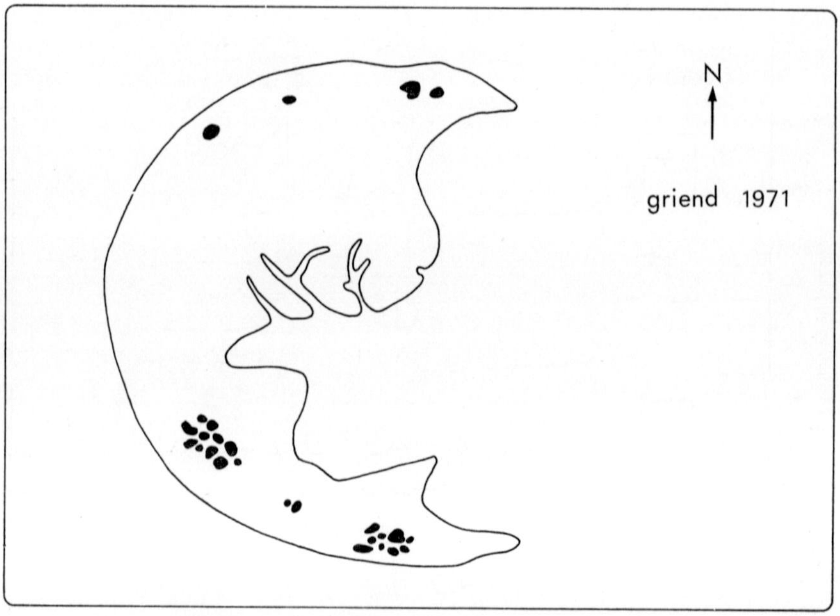

Fig. 7. The distribution of nesting sandwich terns in groups and colonies on Griend in 1971. From North to South 6 colonies can be distinguished, consisting of 2, 1, 1, 11, 2 and 9 groups respectively.

gives the spatial distribution of groups and colonies on the island in 1971. The situation presented should be regarded as representative for all years. From North to South six colonies can be distinguished of 2, 1, 1, 11, 2 and 9 groups, respectively.

The temporal distribution of egg-laying of the 1971 colonies (Fig. 6) shows that egg-laying was markedly synchronized for the larger colonies, but not for the smaller ones.

The total number of breeding pairs present on the island has been termed the island population. Figure 8 gives for five years of observation its distribution of egg-laying. In all years there was a marked peak of laying in the month of May, followed by a second peak 4-5 weeks later. The identification of ringed breeding birds has shown that these 'first' and 'second' peaks of egg-laying referred to recidivist [1] and recruit breeders, respectively. In 1970, between the first and second peak, another one was observed. This unusual peak of laying probably referred to a group of breeding birds which made an unsuccessful breeding attempt in another area after which they settled on Griend.

2.3. SITE SELECTION

The sandwich terns which visited the island in the pre-laying period always concentrated on the ridge of dunes of the island. However, on this ridge the size and the position of a particular group might change considerably from day to day or even from hour to hour. Very often, small groups united with larger ones. Less frequently, large groups splitt up. Sometimes, places which had been occupied for several days, were suddenly deserted, while on other occasions large groups steadily moved in a certain direction causing a slow but gradual change of position over a period of several days. The tendency to move from one place to another seemed at least partly related with the fact that sandwich terns are very sensitive to disturbance. In 1970, for instance, the terns immediately concentrated on new places after a high tide had flooded the whole island on April 29. On other occasions new places were chosen in reaction to human disturbance. However, it should be emphasized that the terns very often changed their position on the ridge without any obvious external cause.

In the beginning of May, when groups of terns suddenly appeared on the island in day-time, the marked mobility of the birds had suddenly vanished. For the first time individual pairs defended a territory and within a few days of their appearance eggs were laid. In succession several groups settled on the island with short intervals. They usually occupied places which had been frequently visited by concentrations of breeding birds during the evening. However, sometimes new places

[1] Recidivist breeders I have called birds which have at least bred once before. I have used this term (borrowed from fisheries-biology) instead of 'experienced breeder' which is usual in the ornithological literature. In the context in which I use the word recidivist, experienced would be confusing since it stresses only one out of several possible causes of differences in reproductive success between first-time breeders and birds which have bred before.

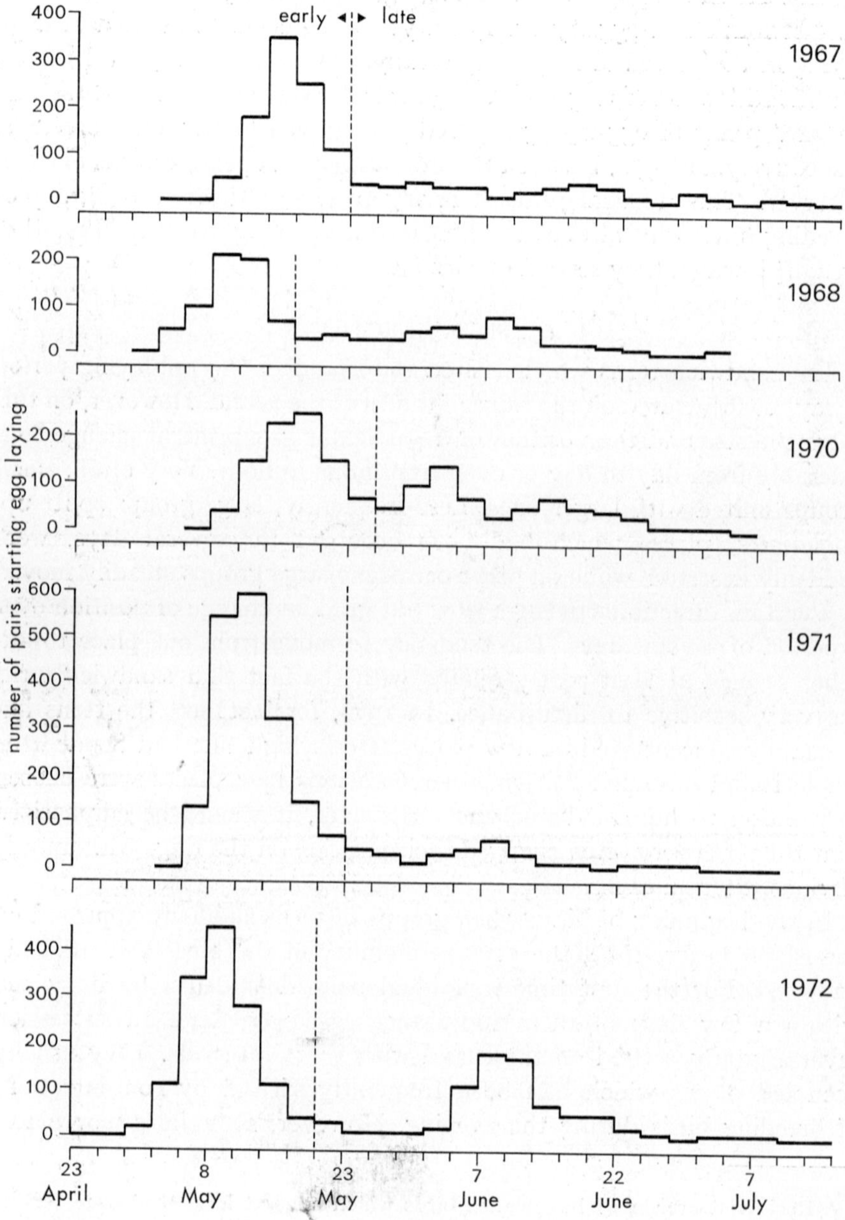

Fig. 8. Distribution of egg-laying in the Griend population of sandwich terns in five years of observation.

Plate 1. Aerial view of Griend showing the outlying sandy ridge and the marshy inner part of the island.

2

3

were chosen. Among breeding birds which settled late in the season this seemed to be even a rule.

The behaviour of the sandwich tern in the pre-laying period indicates that, until a few days before the start of egg-laying, site-attachment is either absent or poorly developed. It was difficult to establish at what time site selection takes place and what plays a role in its determination. According to the characteristics of a large number of colonies studied, the following factors seemed of great importance: (1) vegetation and substrate structure, (2) height in relation to the sea level and (3) the presence of other Larids.

Vegetation and substrate structure.

In all years of observation the places selected for nesting were strikingly similar with respect to vegetation and substrate structure. There was a strong preference for bare sandy places at the borders of vegetation. Although the area chosen might be partly covered with shells, the terns clearly avoided an area where shells were abundant.

In April-May, when the majority of terns settled, plenty of sites were available. However, as a consequence of vegetation development, the area suitable for nesting decreased rapidly in the course of the season.

The presence of nesting sandwich terns importantly influences vegetation development; the growth of plants present within the colony area is strongly retarded by the many droppings of the nesting birds as a result of which the area remains bare and open. However, the influence of the nesting birds on plant growth strongly depends on the phase of development of vegetation. In the Griend situation the late breeders usually started egg-laying within a period in which the vegetation was in an advanced stage of development and the faeces produced by these birds never had a similar restraining influence on vegetation growth as in the early breeders. Besides a relatively small influence of the faeces of the late breeders on vegetation development, the nesting period of these birds also coincided with the growth period of *Chamomilla maritima* and *Atriplex littorale.* As a consequence of both factors, the colonies of the late breeders were sometimes overgrown with vegetation, which often caused the birds to desert their clutches.

Height in relation to sea level.

Of the various places suitable for nesting as far as vegetation and substrate structure are concerned, the terns always selected the highest ones. However, since the number of available sites strongly decreased

Plate 2. Part of the island seen from the NE (summer 1968). (Phot. J. van de Kam).
Plate 3. Sandwich terns and oystercatchers on the highest part of the island during an exceptionally high spring tide (29 April 1970; wind NW, force 9).

in the course of the season—they were either occupied by the early nesting birds or overgrown by vegetation—the late breeders were very limited in their choice. Although they also tended to select the highest places, they had no other choice left than to settle in a much lower area than the early breeding birds (Fig. 9).

Griend varies in height from 0.0 to 1.5 m above the line of mean high tide. Occasionally parts of the island are flooded when long-lasting strong

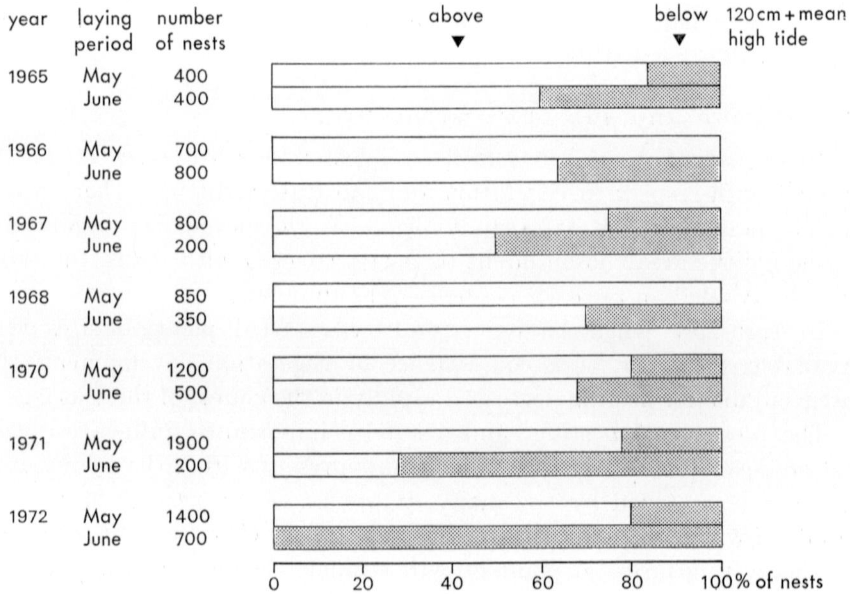

Fig. 9. Habitat selection of early (May) and late (June) nesting sandwich terns. The white area represents the proportion of nests situated at more than 120 cm above the level of mean high tide.

winds cause a rise of water height (Plate 3). For the sandwich tern this means that eggs and chicks are in danger of being washed away. However, the chance to be washed away strongly decreases with increase of the relative height of the nesting area and the highest places are the safest ones. In two years of observation (1966 and 1967) 17.8% and 1.8%, respectively, of the total number of eggs laid were washed away by high tides. In all cases these eggs belonged to breeding birds which had settled relatively late in the season.

The presence of other larid species.

Figure 10 gives the distribution of nests of sandwich terns, arctic terns, common terns and black-headed gulls on Griend in 1968. The situation presented may be regarded as representative for all years of observation. It shows, that the arctic terns mainly nested along the beach side of the island and on the eastern tips. The common terns clearly preferred the more vegetated parts of the island; they usually nested on the higher parts of the salt marsh and the ridge of dunes. The black-headed gulls occupied the most densely vegetated parts of the island; they nested in the salt marsh as well as on the ridge.

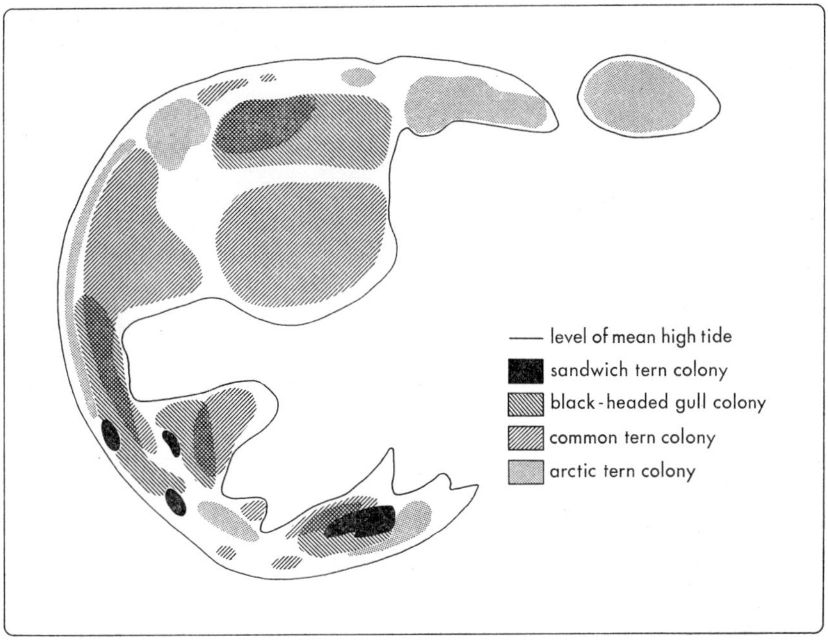

Fig. 10. Distribution of nests of sandwich tern, black-headed gull, common tern and arctic tern in 1968 on Griend.

Together, the nests of arctic terns, common terns and black-headed gulls roughly covered the whole island. As a consequence, the sandwich terns always nested at relatively short distance of either one of these species.

The position of sandwich tern nests in relation to those of the other species seemed largerly dependent on vegetation density. When the terns had settled on open places surrounded by dense vegetation, they were usually surrounded by black-headed gulls (Fig. 11). When the terns occupied a sparsely and irregularly vegetated area they often had settled among the nests of the other species, thus giving rise to mixed colonies (Fig. 12). Sometimes, common tern or black-headed gull nests

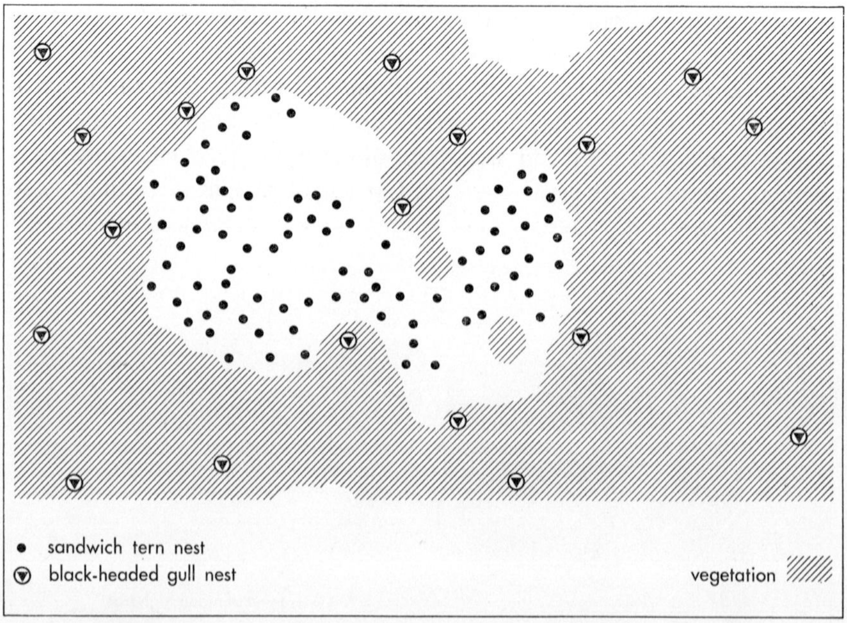

Fig. 11. Example of a mixed sandwich tern–black-headed gull colony. The sandwich terns are encircled by black-headed gulls (colony West, 1970).

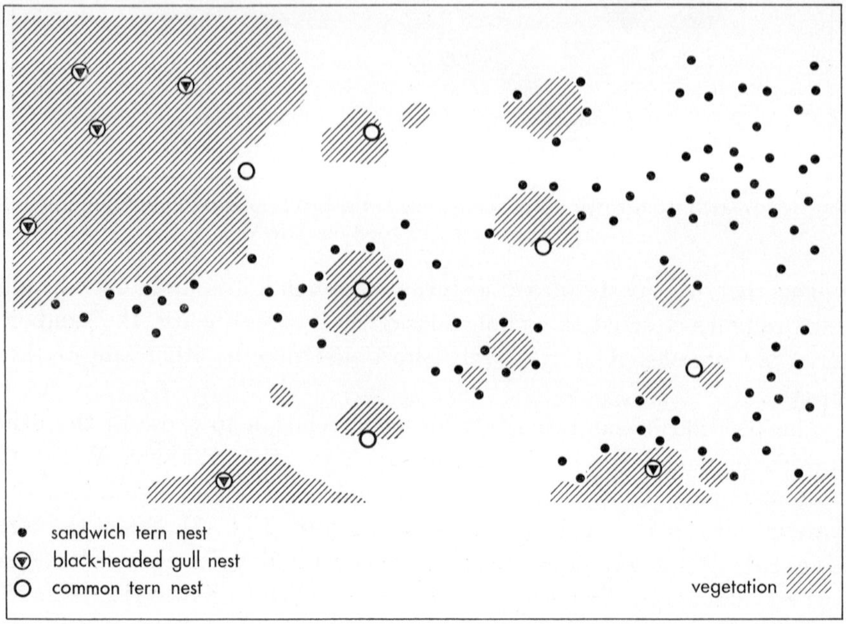

Fig. 12. Example of a mixed sandwich tern–common tern–black-headed gull colony. Several common terns are enclosed by sandwich terns (colony West, 1967).

were enclosed by sandwich terns at all sides. When the surrounding terns were at a very short distance (minimum inter-nest distances of less than 50 cm have been recorded) this often caused the common terns and black-headed gulls to desert their nests.

Since in all years of observation the sandwich terns on the island had no other choice than to settle in the neighbourhood of arctic terns, common terns or black-headed gulls, the Griend situation seemed unable to tell us whether or not the presence of other larids is a factor determining site selection in the sandwich tern. Therefore, data have been collected about

TABLE 2

Associative nesting of sandwich terns with black-headed gulls and common (or arctic) terns in the Netherlands in the 20th century

Colony	Number of years in which breeding of sandwich terns was established	Occurrence of breeding birds		
		Sandwich tern	Black-headed gull	Common tern
Rottumeroog	21	4	–	(3)
Griend	62	4	3	4
Texel schorren	27	4	4	3
A Makkumerwaard	15	3	3	3
De Beer	25	4	4	4
Scheelhoek	12	4	4	4
Schouwen	39	3	3	2
Texel "De Staart"	7	2	3	1
Texel "Burdets Hop"	6	3	3	1
B Mokkebank	7	3	4	3
Kwade Hoek	6	3	4	(1)
Hompelvoet	8	3	3	2
Rottumerplaat	3	3	–	3
Schiermonnikoog	4	2	–	3
Terschelling	1	1	–	–
Texel "De Bol"	1	2	1	2
Callantsoog	1	3	+	+
Veluwemeer	3	2	4	4
C Oost Flevoland	2	2	3	3
Egb. Petruspolder	1	3	3	2
Goudplaat	3	3	4	3
Kaloot	1	2	3	2
Kallandpolder	2	2	4	3
Braakman	1	3	(3)	3
Saaftinge	2	1	+	+

Occurrence of breeding birds (on average): 1 = 1-10 breeding pairs, 2 = 11-100 breeding pairs, 3 = 101-1000 breeding pairs, 4 = 1001-10000 breeding pairs, + = present but numbers unknown, – = not present, brackets indicate breeding birds present, but at distances greater than 100 m.

other sandwich tern colonies in the Netherlands. The results are summarised in Table 2. A distance of one hundred meters between the nearest nests of the different species was taken as a criterion to decide whether or not the species concerned nested in association. As many observers have not distinguished between common terns and the rather uncommon arctic terns, both species are taken together as 'common terns'. The colonies are arranged in order of the number of years in which breeding was established. Three groups are distinguished: (A) colonies occupied for more than 10 years. They are the large 'classical' colonies, usually occupied for many years in succession. (B) Colonies occupied for 6-10 years. (C) Colonies occupied for 1-5 years.

It follows from the table that in the Netherlands sandwich terns nested together with black-headed gulls and with common terns in 20 and 22 colonies, respectively, out of a total of 25. In 19 colonies the sandwich terns nested together with both black-headed gulls and common terns. With respect to associative nesting the groups of colonies distinguished are not different.

In contrast to the Griend situation, in other colonies only a relatively small part of the total area was occupied by the nesting terns and gulls. In general, for each of the species mentioned plenty of sites were available which to the human eye seemed suitable for nesting. This suggests that associated nesting of sandwich terns with arctic terns, common terns and black-headed gulls is not a question of chance but that intra-specific attraction brings the various species together.

2.4. FOOD

2.4.1. Food and feeding habits.

Data from the Griend colony on the food of the sandwich tern are given in Table 3. It shows that the fish fed to the partners (pre-laying and incubation period) as well as to the chicks almost exclusively consisted of *Ammodytes* and *Clupea*. *A. lancea* and *A. lanceolata* on the one hand and *C. harengus* and *C. sprattus* on the other have been taken together since they could not be distinguished in the field. Data on fish collected near the nests in the post-hatching period indicate that *A. lancea* was fed much often than *A. lanceolatus*. The ratio of *C. harengus* to *C. sprattus* appeared to vary throughout the season and between different years. Ammodytidae and Clupeidae have been mentioned as the main food of the sandwich tern throughout its range in Western Europe (DIRCKSEN, 1932; ROOTH, 1965). Observations in the Griend colony showed that prey species other than *Ammodytes* and *Clupea* nearly always were refused by the chicks. This indicates that the choice of food of the sandwich tern is not primarily determined by food availability, but by high specialisation.

The sandwich terns of the Griend colony mainly catch their food off the North Sea coast of Vlieland and Terschelling and the channel in between (Figure 1). The prey is caught by plunge-diving. *Ammodytes lancea* and *lanceolata* are caught in places with strong currents where, as a result of water turbulence, these bottom-living species come to the

TABLE 3

Frequency of occurrence of different fish species in the food of the sandwhich tern in %

	Fish fed to the partner 1966	Fish fed to the chicks 1966	1967	1970
Total number of fish	468	2351	395	356
Ammodytes lancea/lanceolata	52.2	30.8	35.5	64.3
Clupea harengus/sprattus	44.0	66.5	62.8	34.0
Osmerus eperlanus		0.6	0.8	
Gadus merlangus		0.1		
Zoarces viviparus		0.1		
Gobius spec.		0.6	0.3	0.3
Sygnathus spec.	0.2	0.1	0.2	
Gasterosteus aculeatus			0.3	
Platichthys flexus	0.6	0.1		0.6
Unknown	3.0	1.1	0.1	0.8

water surface. Clupeids often move through the area in relation with the tide and, occasionally, large numbers may come within reach of the terns in shallow water. Clupeids may also come within reach of the terns when shoals are driven to the water surface by predatory fish (mainly mackerel).

With respect to the distribution of the fish within the area the hunting behaviour of the terns is of interest. In searching for prey, the terns tend to spread over the feeding area. However, they strongly react to each other's behaviour and immediately join a conspecific fishing successfully. In this way hundreds of terns may come together if a rich source of food is detected. This behaviour promotes that a large number of terns can benefit from a discovery made by single birds and seems a very successful adaptation to prey which is only locally abundant.

2.4.2. Food supply and weather conditions.

In the Griend situation the distance between the colony and the main feeding area measures some 15 to 25 km, which means that each fishing trip needs at least half an hour to an hour flying. This might importantly limit the number of fish supplied for the chicks. In addition, the plunge-diving method of catching fish is known to be highly sensitive to the weather. For this reason it seemed interesting to assess the amount of food fed to the chicks under different weather conditions.

Methods.

In 1966 groups of nests with chicks of known age were selected for detailed study. In each nest to be watched only one chick was present. On a number of successive days a group of nests was continuously watched from 3.30 a.m. until 21.30 p.m. (if possible the chicks were followed from hatching until leaving the colony). Observations were made as follows: On arrival fishcarrying adults were identified in relation to their nest and thereupon followed until disappearance of the fish. For each fish, the final destiny, the species and the estimated length were recorded. Length was estimated in relation to the known average length of the sandwich tern's bill. Before starting the observation estimating length was practised by placing a large number of fish of various length into the bill of a dead sandwich tern. A comparison of the estimated and the measured length of these fish showed that this training highly improved the accuracy of the length assessments. For the field observations it can be held at $\pm$ 10%. The relation between length and weight of fish collected in the colony and on fishing boats around the island allowed us to transform numbers into weight afterwards (Fig. 14).

Results.

Figure 13a gives the number of fish fed daily under normal weather conditions (wind-speed less than 20 miles/hour) to chicks varying in age between 1 and 5 days. The figure indicates that the number of fish fed daily did not change with the age of the chicks. However, the size of the fish supplied gradually increased with increase of age. For instance, one day old chicks were mainly fed on clupeids up to 5 cm, whereas four day old chicks were fed on clupeids up to 12 cm and ammodytids of 8 to 14 cm. Apparently, in the sandwich tern the demand for fish of the growing chick is fulfilled by supplying increasingly bigger fish.

In 1966 a 5-day period of stormy weather (daily average wind-speed more than 20 miles/hour) was recorded between 24 and 29 June. The number of fish fed daily in this period to chicks varying in age between 1 and 5 days is given in figure 13b, showing that the number of fish supplied per chick strongly decreased with increase of age. The marked decrease of the number of relatively big fish fed to the chicks in stormy weather seems related to at least two different factors:

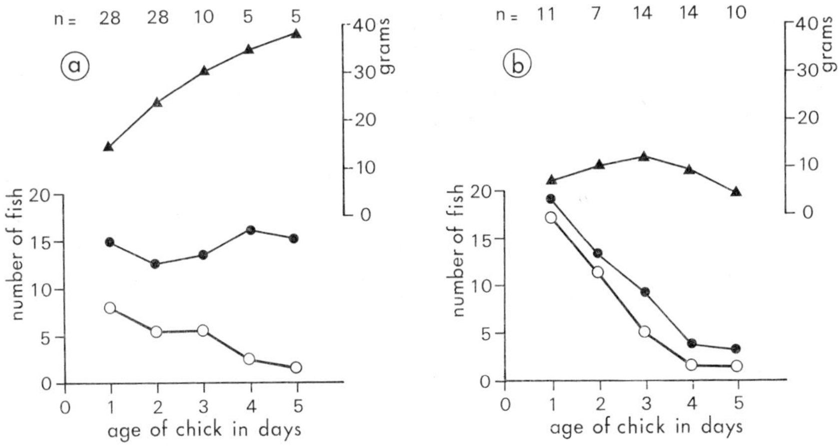

Fig. 13. Number and weight of fish fed daily to chicks varying in age between 1 and 5 days, (a) under normal weather conditions (wind-speed less than 20 miles/hour) and (b) in stormy weather (wind-speed more than 20 miles/hour). Triangles = weight of fish; black dots = total number of fish; open dots = number of small fish. The number of observations (n) is given in chickdays.

(1) A varying proportion of the fish caught by the sandwich tern is stolen by black-headed gulls. They usually steal the bigger fishes and in stormy weather the proportion of fish stolen strongly increases (see II.2.2.6.).

(2) Within the stormy period relatively big fish became more difficult to catch. The bigger fishes are mainly caught off the North Sea coast of Vlieland and Terschelling, whereas smaller fish is also caught on the tidal flats in the neighbourhood of the island. Between 24 and 29 June we noticed that the parents of 1 and 2 day old chicks mainly foraged in shallow water in the vicinity of the island; they were rather successful in catching small fish. The parents of the older chicks persisted in visiting the feeding grounds off the coast of Vlieland and Terschelling. They usually returned to the colony without food. Apparently the parents of the older chicks were fixed upon capturing relatively big fish at the out-laying feeding grounds; ignoring the presence of small fish present in the neighbourhood of the island.

To examine the consequences of strong winds on the food intake of the chicks numbers have been converted into weight. It has been mentioned before that the field observations did not allow us to discriminate between *C. sprattus* and *C. harengus* and between *A. lancea* and *A. lanceolata*. Information on the occurrence of each species within the food of the terns had to be based on finds of fish in the colony, which gave the following results: *C. sparattus* 28, *C. harengus* none, *A.*

lancea 12, *A. lanceolatus* 2. Since all Clupeidae turned out to be *C. sprattus*, the upper graph in Figure 14 can be taken to refer only to this species. It is possible of course, that *C. harengus* was supplied for the chicks but not recorded in the finds. However, within the range of lengths relevant to our data, the length-weight relationship of C. *harengus* differed only slightly from that of *C. sprattus* and any important influence of the possible occurrence of *C. harengus* on the final results of weight can be excluded.

A. lancea and *A. lanceolata* both occurred in the finds and both have been included in the lower graph of Figure 14. The number of specimen found does not allow any conclusion with respect to their relative occurrence in the food of terns, but the small difference in the relevant length/weight relationship of both species makes it unlikely that this has influenced our results on weight importantly.

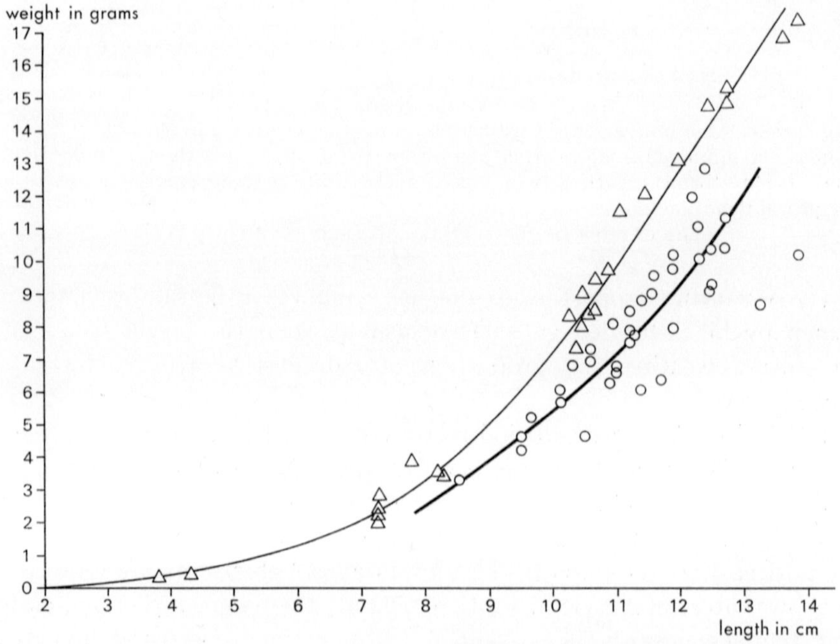

Fig. 14. Relation between length and weight of *Clupea sprattus* (triangles), *Ammodytes lancea* and *A. lanceolata* (circles) collected in the colonies and near the island in June 1966.

The daily weight of fish fed to chicks of different age under normal and stormy conditions are represented by the upper graphs in the Figures 13a and 13b. It shows that under normal conditions the amount of fish fed strongly increased with increase of age. In stormy weather the food intake of the chicks was much less than under normal conditions and the older chicks suffered more heavily from food shortage than the younger ones.

Data on the influence of food shortage on chick mortality is available from frequent examination of chicks found dead in the colony.

Figure 15 gives the cumulative percentage of chicks hatched and found dead throughout the 1966 season. It follows that a high proportion of chicks (88 out of 164) was found dead during or immediately after the 5-day period of stormy weather between 24 and 29 June. All the chicks found had considerably lost in weight in the given period; 5-day old chicks dying on 29 June even weighed less than on hatching and we have no reason to doubt that chick mortality between 24 and 29 June was directly caused by food shortage.

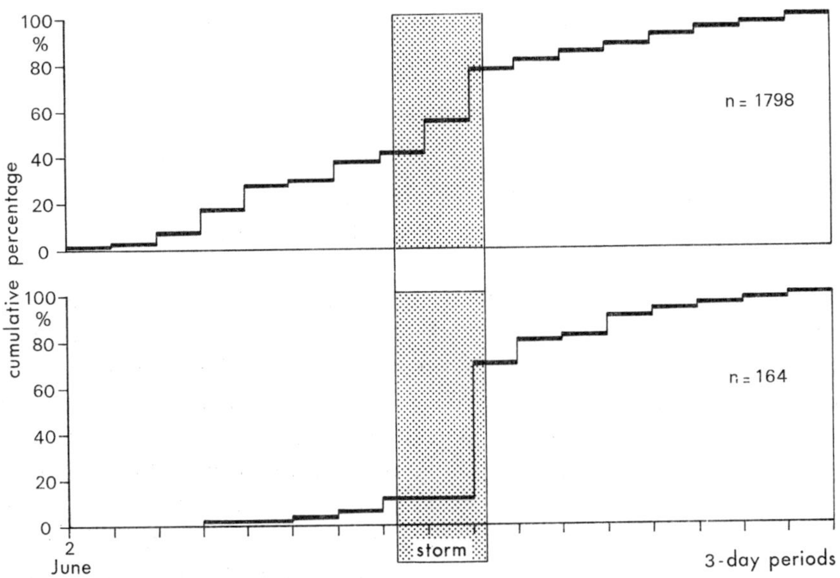

Fig. 15. The cumulative percentage of chicks found hatching (upper graph) and found dead (lower graph) throughout the 1966 season.

A 5-day period of stormy weather, as recorded in 1966, seems rather unusual in the hatching period of sandwich tern chicks. In all years of observation there were occasional periods of food shortage caused by stormy weather, but such periods did not generally exceed two or three days. I often got the impression that the tern chicks were capable of surviving a period of two to three days with hardly any food. Such a strong correlation between chick mortality and weather conditions as in 1966 was not recorded in other years of observation. In fact many chicks were found dead throughout the season in all years. The majority of these chicks, however, were second-hatched chicks from 2-egg clutches. Such chicks are easily deserted by the parents at the moment of nest leaving. Since it was not possible to determine whether food shortage or desertion was the primary cause of death of chicks which had died of starvation, it remains uncertain if and to what extent food shortage caused chick mortality between 1967 and 1972.

Summarizing we conclude that the sandwich tern is dependent on only a few prey species, which are caught at half an hour to an hour flying from the colony. Moreover, the plunge-diving method of catching fish is known to be highly sensitive for adverse weather conditions, *e.g.* storm and fog (HAWKSLEY, 1957). In 1966 a large number of chicks died as a consequence of food shortage.

2.4.3. Food abundance and clutch size.

LACK (1947, 1948, 1954, 1968) has drawn attention to a number of general trends in clutch-size, size of eggs and incubation of seabirds, which are of interest with respect to the food-egg relationship in the sandwich tern.

According to LACK, the normal clutch of a species corresponds to the maximum number of young that the parents can successfully raise. In the sandwich tern 1 or 2 eggs are laid. This is the smallest clutch in terns of the northern hemisphere. I never had an indication that any pair of the sandwich tern population on Griend raised more than one chick between 1965 and 1969. In 1970, 1971 and 1972, which seemed to be very favourable years as far as food is concerned, the proportion of 2-egg possessing pairs raising two chicks was still very small. Therefore, it is very likely that in case of the sandwich tern a clutch of two is indeed close to the limit set by food for the young when food is plentiful.

There is a general tendency for eggs to be proportionally smaller in species with larger clutches. However, in spite of the small clutch, the size of the egg of the sandwich tern is also relatively small. This suggests that food is sparse for the female around the time of laying. Moreover, the sandwich tern has the unusually long laying interval of 2 to 5 days, which seems an adaptation to the possibility of food-scarcity in the period prior to egg-laying.

In the sandwich tern incubation starts with the first egg and the young hatch asynchronously. This phenomenon is rather common in seabird species which feed at a long distance from the nest and which periodically have to cope with food shortage. As a consequence of a relatively long period between the hatching of the chicks, the smaller one quickly starves when food gets short. According to LACK this is an adaptation for reducing the broodsize to that which the parents can effectively feed, without food being wasted on chicks that would die anyway.

2.4.4. Food abundance and seasonal timing of egg-laying.

In all years of observation the sandwich terns of the Griend population laid their eggs highly synchronized in the second half of May. This seasonal timing of egg-laying might be adapted to feeding conditions.

In the North Sea area *Ammodytes lancea* and *A. lanceolata* as well as *Clupea harengus* and *C. sprattus* spend the winter in deeper water (MUUS, 1966). In spring they migrate to the more shallow coastal waters where they may remain till the end of the summer. Unfortunately data on the occurrence throughout the season of Ammodytidae and Clupeidae in the Wadden Sea area are scarce. The catches of shrimp fishing boats of Nieder-Sachsen and Schleswig Holstein in the German part of the Wadden Sea (MEYER-WAARDEN & TIEWS, 1965; TIEWS, 1971) suggest that the occurrence of *Ammodytes* spec., *Clupea harengus* and *C. sprattus* may vary considerably from place to place and from year to year. However, it should be noted that the data concerned refer to catches by bottomscraping nets. These catches may originate from various depths. Therefore, for an understanding of the food availability of the surface feeding sandwich tern the results may be misleading.

POSTUMA, ZIJLSTRA & DAS (1965) have shown that large numbers of *C. harengus* larvae measuring 2.5 and 3.5 cm enter the Dutch Wadden Sea between February and April. These larvae grow rapidly, reaching a length of 3 to 5 cm in June (ZIJLSTRA, pers.comm.). As mentioned before, in the Griend colony the newly hatched chick of the sandwich tern is almost exclusively fed on very small clupeids. Bigger fish is usually refused by these chicks. Thus, in the hatching period of the chicks the availability of small fish seems an important condition to be fulfilled. Therefore, the timing of egg-laying of the sandwich tern population on Griend might be related to the presence of large numbers of small *C. harengus*.

PART II

ASPECTS OF FUNCTION

1. NEST DISTRIBUTION AND REPRODUCTIVE SUCCESS

1.1. INTRODUCTION

The present chapter deals with the possible influence on reproductive success of nest distribution in the sandwich tern. Although reproductive success is determined by the number of offspring reaching fertility, my observations were confined to the life span of the progeny in the breeding season, as chicks could not be followed after fledging.

To demonstrate the influence on success of a particular behaviour pattern, one might compare under similar conditions individuals which do and which do not possess this pattern. In general, however, it is very difficult or even impossible to select from the population individuals in which a particular behaviour pattern is completely lacking. It is, therefore, usually more practical to study the entire range of variation of the behaviour pattern among the individuals of the population. In my study this amounts to examining clutch size, laying date, hatching success and fledging rate in relation to the inter-nest distances and nest configuration in the whole range of the island population, colonies and groups as distinguished in the preceding chapter.

1.2. METHODS

1.2.1. Approach to the analysis.

In the previous part, chapter 1.2.2., with respect to the distribution of nests, the following units have been distinguished: groups — colonies — island population. Each of these units may be characterised by aspects of size, density and synchronization as given below (the terms in italics are used in the following sections).

For a better understanding of the possible significance for success of nest distribution in the sandwich tern, I have tried to relate each of the above mentioned characters independently of each other to success. To this end, throughout the season in different years of observation, groups of nests from different colonies have been selected for detailed

	Size	Density	Synchronization
Group	*group size* (number of nests of the group)	*nest position* (inter-nest distances within the group)	*group synchronization* (synchronization of egg-laying with the other nests of the group)
Colony	*colony size* (number of nests of the colony)	*group position* (inter-group distances within the colony)	*colony synchronization* (synchronization of egg-laying with the other nests of the colony)
Population	*population size* (number of nests of the population)	*colony position* (inter-colony distances within the population)	*island synchronization* (synchronization of egg-laying with the other nests of the population)

study. The success of these nests was determined and thereupon related with the various characters of the nest distribution pattern.

Unfortunately, some of the characters mentioned, viz. population size, colony position, group position and colony synchronization, could not be incorporated in the analysis. Leaving out of consideration some of the characters may have influenced the results on the others.

The relationship of population size and colony position with success can only be studied if different populations can be compared under similar environmental conditions. My data refer to one population. I could have compared my data from one year to the other. However, as environmental conditions are likely to vary between years, such a comparison has not been made. Leaving out of consideration population size and colony position cannot have influenced the other relationships studied.

In general, groups varied considerably with respect to size and shape. However, important differences in group position within the colony did not occur among the colonies studied in 1970, 1971 and 1972. Consequently, I have refrained from incorporating group position in the analysis, the determination of which would, moreover, have caused much disturbance.

Since the temporal distribution of egg-laying of the island population was the result of the distribution of egg-laying in a relatively small number of colonies (Fig. 6), the possible influence on success of colony synchronization and island synchronization could not be separated from each other. Thus, for practical reasons it seemed necessary to choose one unit for analysis. Island synchronization has been chosen. However, in the following pages the close relationship of this character with colony synchronization should be borne in mind.

In addition to the various characters of the nest distribution pattern some other factors, viz. clutch size, date of egg-laying and age, have also been included in the analysis. In succession the following factors will be related to success in the coming sections: (1) clutch size, (2) date of egg-laying, (3) group synchronization, (4) nest position, (5) group size, (6) colony size, (7) island synchronization, (8) age.

1.2.2. Measurement of success.

To determine reproductive success I have collected per nest the following data: (1) number of eggs; (2) laying date per egg; (3) egg mortality; (4) date of hatching per egg; (5) fate of the chicks (fledged, died, if possible cause of death).

Each nest to be observed was marked with small inconspicuously coloured wooden pegs. Incubating birds did not noticeably react to these markers. Neither do we have indications that the markers influenced the behaviour of predators (*e.g.* scared them), but the absence of such an effect is far more difficult to establish. However, even if it was present it is likely to be the same for all nests.

Up to 1967 we checked the nests daily. However, as human disturbance sometimes influences predation—for instance, black-headed gulls and oystercatchers robbed eggs during our visits and one black-headed gull even seemed to have specialised on it—from 1968 onwards we avoided this interference as much as possible by checking nests every three days and by speeding up the proceedings through the use of a taperecorder instead of pencil and notebook. Observations from a hide, before, during and after this disturbance, showed that in this way our influence on predation was minimal. During the egg-laying and hatching periods, however, daily checking was necessary and a slight effect of this increased interference cannot be excluded.

Eggs were regarded to have hatched if the chick had fully emerged from the shell. Eggs remaining unhatched beyond the maximum incubation period were examined and classed as addled, died in hatching, *etc.*

A considerable number of eggs disappeared in the hatching period. The majority of these eggs belonged to two-egg clutches discarded by the parents when leaving with the first-born chick, usually as a consequence of either disturbance or hatching failure. Such eggs were nearly always taken by predators; thus the primary cause of death usually remained unknown. For this reason these cases were taken together and classified as 'lost during hatching'.

Theoretically our data on clutch size and egg mortality could have been affected by eggs laid and disappeared within the timespan between two visits. However, the only egg predators on Griend, black-headed gull and oystercatcher always leave egg-remains close to the nest, because they are unable to take away intact eggs. Therefore, not many cases of egg-robbing can have been overlooked.

Chicks were banded immediately after hatching. We always used numbered aluminium rings, but sometimes celluloid colour rings were used in addition to make individual birds recognizable in the field. During the whole season dead chicks were collected and examined on the cause of death. Mortality percentages were never found to be different between banded and non-banded chicks, which

Plate 4. Sandwich terns concentrated in the nesting area in the pre-laying period. (phot. J. van de Kam).

indicates that banding did not influence chick viability. When 25-30 days old, all chicks normally fly and consequently the chicks disappeared from our data. For practical reasons, chicks recorded to be alive at the age of over 14 days, were regarded to have reached the fledging age. This assumption seems justified, because mortality has very strongly decreased at this age.

During the first two weeks after hatching chicks were left undisturbed as much as possible. To establish the number of chicks after this period and until the chicks at the age of four weeks were able to fly, the entire island was censussed with intervals of three days. The majority of the chicks hide in dense vegetation. They could be caught easily by searching the vegetation on the island. Many chicks, however, leave the vegetation and concentrate in large groups on the beach when they are about to fly. This category was successfully caught by driving them into a 50 m long wire netting fike with a chamber at the end. In this way all age classes had about equal chances to be recovered.

On the average 50% of the chicks present on the island was caught each time, and as chicks could maximally be caught four times before they fledged at the age of four weeks, we must have recovered more than 90% of the chicks produced in the population.

In the above described way nests have been checked for success in 1965, 1966, 1967, 1968, 1970, 1971 and 1972. Between 1965 and 1968, however, only hatching success was determined. Moreover, between 1965 and 1967 human disturbance and in 1965 poisoning by organochlorine pesticides may have influenced the results obtained. Therefore, the results given below will exclusively refer to the data collected between 1970 and 1972. One exception has been made in the section on colony size, where the 1965-1967 data have been used in addition to the 1970-1972 data.

1.3. RESULTS

In the following sections success will be expressed as:
Hatching success (= the percentage of eggs hatching);
Fledging success (= the percentage of hatched chicks which result in fledged young);
Breeding success (= the percentage of eggs giving rise to fledglings).

Each of these expressions of success incorporates a possible influence of clutch size on success. Since clutch size varied between nests, it was necessary first to study the relation between clutch size and success before correlating success with the various aspects of colony breeding.

1.3.1. Clutch size and success.

Sandwich terns normally lay a clutch of one or two eggs. Although a nest may contain three eggs, this is rare and it is likely that such eggs originate from more than one female.

Only 1- and 2-egg clutches were studied and compared for success in 1970, 1971 and 1972. The daily nest checks made it possible to distinguish real 1-egg clutches (the only 1-egg clutches taken into account) from 2-egg

Plate 5. A sandwich tern colony situated at very short distance from a black-headed gull colony (Kwade Hoek, June 1970). The terns are in the middle of the photograph; the nesting gulls are visible in the background and a single individual in the foreground. (phot. J. van de Kam).

clutches from which one egg had been robbed. To avoid season-correlated influences within our data (*e.g.* clutch size seemed to decrease in the course of the season) nests were selected from a period in which little or no variation in clutch size occurred.

Table 4 gives the breeding results of 193 and 233 nests in 1970 and 1971, respectively. In both years hatching success was not statistically different between 1- and 2-egg nests (X^2 test: 1970 — $0.20 > p > 0.10$, 1971 — $0.90 > p > 0.80$), but fledging success and breeding success were significantly higher for the 1-egg nests in both years (X^2 test: fledging success 1970 — $p < 0.005$, 1971 — $p < 0.005$; breeding success 1970 — $p < 0.005$, 1971 — $p < 0.005$).

TABLE 4

Breeding results of 1 and 2-egg clutches in 1970 and 1971

	1970 (193 nests)		1971 (233 nests)	
Glutch size	1 egg	2 eggs	1 egg	2 eggs
Eggs laid	74	232	47	372
Eggs hatched	65	186	34	273
Chicks fledged	52	98	30	149
Hatching success	87.8%	80.2%	72.3%	73.5%
Fledging success	80.0%	52.7%	88.2%	54.6%
Breeding success	70.3%	42.2%	63.8%	40.1%

Table 5 gives some more details on chick mortality. There was a marked difference in fledging success between first and second-hatched chicks within a clutch of two (X^2 test: 1970 $p < 0.005$, 1971 $p < 0.005$,

TABLE 5

Fledging success of chicks from 1- and 2-egg clutches in 1970, 1971 and 1972

	Clutch size	Category of chick	Number of chicks: Ringed	Fledged	% fledged
1970	1 egg		214	177	82.7%
	2 eggs	1st hatched chick	829	695	83.8%
		2nd hatched chick	495	99	20.0%
1971	1 egg		182	142	78.0%
	2 eggs	1st hatched chick	1058	848	80.2%
		2nd hatched chick	748	88	11.8%
1972	1 egg		252	108	42.8%
	2 eggs	1st hatched chick	150	74	49.3%
		2nd hatched chick	65	10	15.4%

1972 $p < 0.005$). First-hatched chicks from 2-egg clutches seemed to have slightly better chances to survive than chicks in a clutch of one. Although these differences were not statistically significant for each year separately, if all data are taken together the difference is highly significant (X^2 test: $p < 0.005$).

The high mortality rate of second-hatched chicks in the 2-egg nests suggests that most sandwich terns were not capable of raising two chicks. In fact between 1965 and 1968 we never observed any pair fledging more than one chick (parents feeding more than one chick beyond the age of fledging were never seen). Banding results in 1971 and 1972 showed that a small percentage of breeding birds fledged two chicks: 1971 — 64/861 = 7.4% (eggs laid in May and June); 1972 — 19/100 = 19.0% (May), 2/65 = 3.1% (June); (the percentages are based on nests in which both chicks were banded.) It should be noted, however, that 1971 and 1972 were favourable years in many respects.

We may conclude from these results that hatching success did not, but fledging success and breeding success did differ significantly between 1- and 2-egg clutches in 1970 and 1971. Therefore, in the following sections clutch size has to be taken into account if groups of nests are compared for fledging or breeding success.

1.3.2. Date of egg-laying and success.

In a number of Larid species late breeders, in comparison to earlier ones, have been recorded to have a relatively low breeding success (common tern, AUSTIN, 1945; arctic tern, VEEN, unpublished; black-headed gull, PATTERSON, 1965; kittiwake, COULSON & WHITE, 1958, 1959, 1966). This phenomenon might be caused by differences in the behaviour of early and late breeders as well as by seasonal changes in the pressure of environmental factors. Since it is our aim to analyse the possible influence of behaviour aspects on reproductive success, the influence of seasonal fluctuations in environmental factors should be excluded from this analysis as much as possible. Therefore, it seemed necessary to compare early and late breeders for success before starting our analysis of the relation between the various aspects of the nest distribution pattern and success.

The distribution of egg-laying of the sandwich tern population on Griend has been given in Figure 8. In all years there was a marked peak of egg-laying in the second half of May. In 1970, 1971 and 1972 these early 'peak breeders' have been separated from the late breeders as indicated in the figure. In Table 6 both groups have been compared for success. To avoid the possible influence of clutch size differences within the results, 1- and 2-egg clutches have been separated. The table shows that breeding success was significantly higher for the early peak breeders than for the late breeders (X^2 test, for 1- and 2-egg clutches in all cases: $p < 0.005$).

TABLE 6

Breeding success of early and late breeding birds in 1970, 1971 and 1972. The data refer to the whole island population

		1970		1971		1972	
		Early	Late	Early	Late	Early	Late
1-egg nests	Eggs laid	355	288	294	238	245	436
	Chicks fledged	240	11	163	1	169	82
	Breeding success	67.6%	3.8%	55.4%	0.4%	69.0%	18.8%
2-egg nests	Eggs laid	1740	458	3008	86	2236	500
	Chicks fledged	731	14	1036	2	1160	113
	Breeding success	42.0%	3.1%	34.4%	2.3%	51.9%	22.6%

1.3.3. Nest distribution and success.

This section deals with the possible significance for success of the nest distribution pattern in groups and colonies. For this purpose, those nests that were under regular observation in 1970 and 1971, have been grouped with respect to group synchronization, nest position, group size and colony size and thereupon related to success.

To avoid interference by season-correlated differences between nests as much as possible, nests had to be selected from a rather restricted period. It was decided to take only the peak breeders of the population for analysis, as they started egg-laying in great numbers within a relatively small period. In both years they were selected as follows (see also Fig. 8):

1970 — clutches started between 8/5 and 25/5;
1971 — clutches started between 5/5 and 23/5.

With respect to the measurement of success it is important to remember that all chicks leave their nests within a few days after hatching, mixing with other birds of the island population. This means that the distribution of nests in group and colonies refers to a pattern which disappears shortly after hatching. This pattern can, therefore, only have an influence on success during incubation and the few days between hatching and leaving.

Unfortunately we were not able to determine success at the moment of leaving, as any disappearance of a chick from its nest might be a consequence of either predation or leaving the colony. For this reason the above mentioned aspects of the nest distribution pattern were related to success at the moment of hatching, *i.e.* hatching success.

1.3.3.1. *Group synchronization.*

Figure 16 gives the distribution of egg-laying of 7 groups of sandwich tern nests studied in 1970 and 1971. The figure shows clearly that for each group egg-laying was markedly synchronized. To study the influence of group synchronization on hatching success, each group has been divided in 'peak' and 'tail' breeders as indicated in the figure by a broken line. Both groups have been compared for hatching success. It follows, that for each group hatching success was higher for the 'peak' breeders than for the 'tail' breeders (data summarized: X^2 test, $p < 0.005$).

We conclude from these results that the peak breeders of a group are more successful than the tail breeders. However, the above results cannot tell us if the differences found are caused by group synchronization. Peak and tail breeders might be different with respect to nest position and a possible correlation between nest position and success may have caused the relationship between group synchronization and success.

1.3.3.2. *Nest position.*

The position of the nests in a group in relation to each other will be expressed as:

(a) minimum inter-nest distance,
(b) nest density,
(c) central versus peripheral position.

Minimum inter-nest distance.

Minimum inter-nest distances appeared to be highly uniform in all years of observation. After the breeding season these distances were measured for all nests under observation. Minimum inter-nest distance was determined as the distance to the nearest neighbour on the second day after the start of egg-laying, this being the best representation of a tern's choice of a nesting site in relation to its neighbours.

In Table 7 the nests are grouped with respect to their minimum inter-nest distance and compared for hatching success. To avoid the influence of group synchronization 'peak' and 'tail' breeders are treated separately.

For the 'peak' breeders as well as for the 'tail' breeders hatching suc-

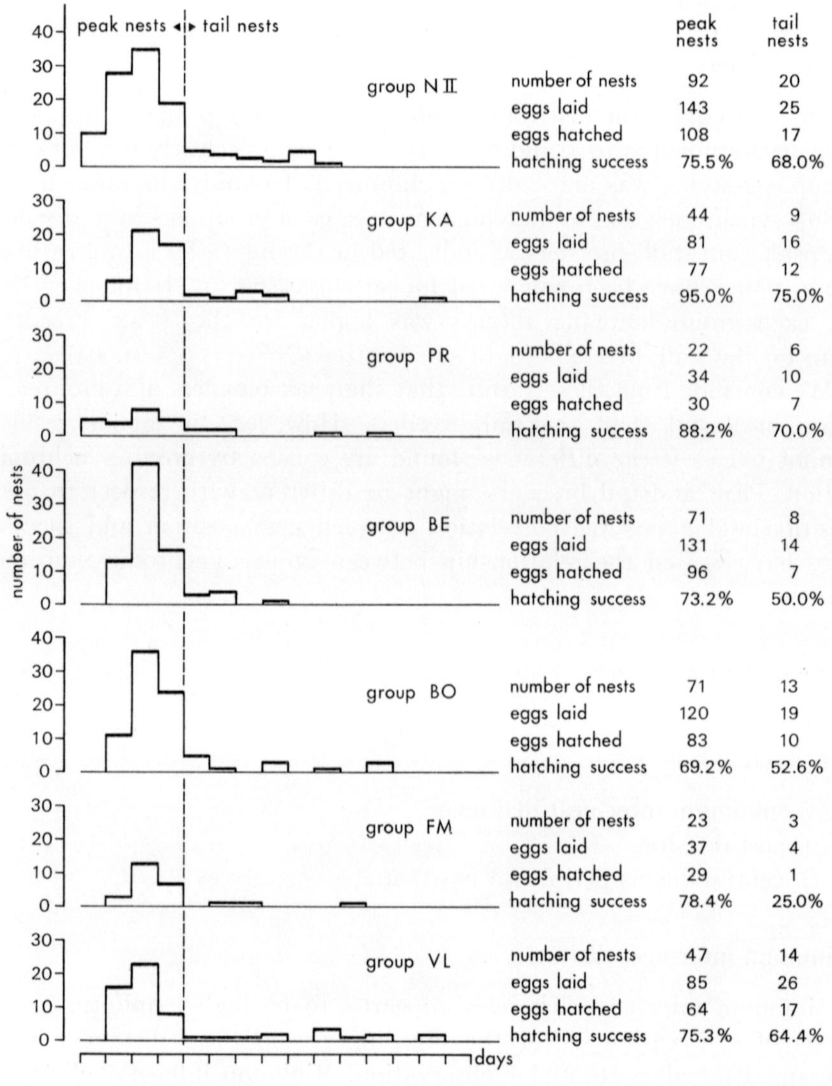

Fig. 16. Distribution of egg-laying on successive days of 7 groups of sandwich terns in 1970 and 1971. The various groups have been divided in 'peak' and 'tail' nests as indicated by the broken line. For each group hatching success is given near the figure.

TABLE 7

Relation between minimum inter-nest distance and hatching success

Peak breeders

Minimum inter-nest distance	Number of nests	Number of eggs laid	Number of eggs hatched	Hatching success
21-30	53	88	61	69.3%
31-40	203	352	291	82.7%
41-50	79	133	97	70.7% (41-100)
51-60	23	35	23	
61-70	5	9	6	
71-80	6	11	8	
81-90	1	2	1	
91-100	1	1	0	
Total	371	631	487	77.2%

Tail breeders

Minimum inter-nest distance	Number of nests	Number of eggs laid	Number of eggs hatched	Hatching success
21-30	8	13	8	61.5%
31-40	35	54	36	66.7%
41-50	17	26	15	57.5% (41-100)
51-60	6	12	7	
61-70				
71-80	2	3	0	
81-90	1	2	1	
91-100	3	4	4	
Total	72	114	71	62.3%

cess was highest for those birds which had laid their eggs at the most common distance (group II against I + III: 'peak' breeders, X^2 test: $p < 0.005$; 'tail' breeders, X^2 test: $p < 0.95$). So, we may conclude that minimum inter-nest distance is related with success apart from group synchronization.

'Peak' and 'tail' breeders were not different with respect to their distribution of minimum inter-nest distances. Consequently, the relationship between group synchronization and hatching success cannot be caused by differences in minimum inter-nest distance.

Nest density.

Inter-nest distance is only one way to express the spacing of nests. In this measure only one neighbour is taken into account and the possible influence on success of neighbouring nests other than the nearest is ignored. In fact each nest present at short distance can be expected to contribute to success. This made it interesting to relate the number of nests within a certain distance, representing for each nest the local nest density, with success.

The majority of birds laid their eggs at a distance of 31-40 cm from the nearest neighbour and the number of nests within a radius of 50 cm, therefore, seemed a good parameter of the density. However, it sometimes changed considerably throughout the incubation period, as a consequence of the disappearance of clutches which were hatched or robbed, or of the appearance of new-comers. Since each nest can be regarded to contribute to density in proportion to the number of days it was present, density was ultimately expressed for each nest as the sum of the number of days on which each neighbouring nest within a radius of 50 cm was present; in short the number of neighbour-days.

The nests studied in 1970 and 1971 were with respect to their density grouped into classes, which then were compared for hatching success. To avoid the influence of group synchronization 'peak' and 'tail' breeders were again treated separately. The results are given in Table 8.

Among the 'peak' breeders hatching success was significantly higher for those birds which nested at the most common density as compared to the higher and lower density groups together (X^2 test: $p < 0.05$). So, density was related with hatching success even when the effect of group synchronization is allowed for.

The 'peak' breeders on the average nested at a higher density than the 'tail' breeders. Consequently, nest density may have influenced the relationship between group synchronization and hatching success. However, if we compare hatching success of 'peak' and 'tail' breeders for each class of density, it follows that 'peak' and 'tail' breeders are different with respect to hatching success independently of nest density (X^2 test on the two lower density classes: $X^2 = 8.73$, df 2, $p<0.025$).

Central or peripheral position within the group.

Besides inter-nest distance and nest density, with respect to nest position another factor can be recognized: position in relation to centre and edge of the group. Although this aspect will be at least partly related to nest density, it might also separately influence success.

In 1970 and 1971 nests checked for success were classified as central (*i.e.* surrounded by neighbouring nests at all sides) or peripheral; both

TABLE 8

Relation between nest density and hatching success. Neighbour-days = the sum of the number of days on which neighbouring nests within a radius of 50 cm are present

Peak breeders

Neighbour-days	Number of nests	Number of eggs laid	Number of eggs hatched	Hatching success
0- 40	129	214	158	73.8%
41- 80	176	300	243	81.0%
81-120	51	89	68	76.4%
121-160	15	28	18	64.3%
Total	371	631	487	77.2%

Tail breeders

Neighbour-days	Number of nests	Number of eggs laid	Number of eggs hatched	Hatching success
0- 40	47	75	46	61.3%
41- 80	24	38	25	65.8%
81-120	1	1	0	0.0%
Total	72	114	71	62.3%

TABLE 9

Central versus peripheral nest position within the group in relation to hatching success

Peak breeders

Nest position	Number of nests	Eggs laid	Eggs hatched	Hatching success
Central	105	189	148	78.3%
Peripheral	266	442	339	76.7%

Tail breeders

Nest position	Number of nests	Eggs laid	Eggs hatched	Hatching success
Central	3	5	5	100.0%
Peripheral	69	109	66	60.6%

groups were related to hatching success. To avoid the influence of group synchronization 'peak' and 'tail' breeders were again treated separately. Table 9 gives the results.

For the 'peak' breeders as well as for the 'tail' breeders a higher percentage of eggs hatched in the central nests than in the peripheral ones. In both cases, however, the differences were not statistically significant (X^2 test: 'peak' breeders — $p < 0.90$; 'tail' breeders — $p < 0.10$).

Table 9 also shows, that the proportion of central nests was higher for the 'peak' breeders than for the 'tail' breeders. Considering that central and peripheral nests were not significantly different with respect to hatching success, their difference in distribution among peak and tail breeders cannot have influenced the relationship between group synchronization and hatching success.

Interrelationships of the different nest-position measures.

The results on minimum inter-nest distance and nest density led us to conclude that both measures of nest position were related to hatching success. However, a large minimum inter-nest distance logically correlates with a low nest density and a short minimum inter-nest distance can be expected to be correlated with a high nest density. The relationship with success of both measures thus might be one and the same. A separation of the influence of both measures was possible in a further analysis of their relation to one another and to success.

Table 10 gives with respect to minimum inter-nest distance and nest density the distribution of the nests studied in 1970 and 1971 (both peak and tail breeders are included). The table shows that the majority of sandwich tern nests were at the same time situated at a minimum inter-nest distance of 31-40 cm and at a nest density of 41-80 neighbour-days.

The same data are summarized in Table 11a. Table 11b gives for these summarized data the corresponding number of eggs laid as well as the number hatched; hatching success is given in Table 11c.

If we consider separately the different classes of nest density, Table IIc shows that nests situated at a minimum inter nest distance of 31-40 cm were in all cases more successful than nests at higher and lower distances. This indicates that minimum inter-nest distance is of influence on success apart from nest density. If we now consider the different classes of minimum inter-nest distance, it follows that nests situated at the most common nest density are not always more successful than nests at higher and lower densities. In fact groups of nests which are of different density but of the same minimum inter-nest distance, are hardly different with respect to success, except for the two groups which are at the upper right and at the lower left hand of the table. It thus appears, that density is of negative influence on success in two situations only: (a) When a low density coincides with a large minimum inter-nest distance and (b) when a high density coincides with a small minimum inter-nest distance.

Finally, the relationship between central-peripheral nest position on the one hand and nest density and minimum inter-nest distance on the other remains to be studied. For this purpose the proportional distribution of central nests in relation to nest density and minimum inter-nest distance is given in Table 11d. (The percentages given in the table are calculated over the different groups of nests as presented in Table 11a). It follows from the table that the proportion of central nests strongly increases with increase of nest density and, although less distinct,

TABLE 10

Nest distribution in relation to minimum inter-nest distance and nest density

Nest density in neighbour-days

Minimum inter-nest distance in cm	0-40	41-80	81-120	121-160	Total
21-30	13	33	10	5	61
31-40	60	134	35	9	238
41-50	55	33	7	1	96
51-60	29				29
61-70	5				5
71-80	8				8
81-90	2				2
91-100	4				4
Total	176	200	52	15	443

TABLE 11

Interrelationships between minimum inter-nest distance, nest density, central vs peripheral nest position and hatching success. For explanation see text

	minimum inter-nest distance in cm.	nest density in neighbour-days 1 - 40	41 - 80	81 - 100	
(a)	21 - 30	13	33	15	number of nests
	31 - 40	60	134	44	
	41 - 100	103	33	8	
(b)	21 - 30	24 (18)	50 (37)	27 (15)	number of eggs laid (hatched)
	31 - 40	96 (77)	232 (188)	78 (61)	
	41 - 100	169 (109)	56 (43)	13 (10)	
(c)	21 - 30	75.0	74.0	55.6	hatching succes (%)
	31 - 40	80.2	81.0	78.2	
	41 - 100	64.5	76.8	76.9	
(d)	21 - 30	7.7	21.2	80.0	nests centrally situated (%)
	31 - 40	7.5	26.8	75.0	
	41 - 100	3.9	25.2	75.0	

with decrease of minimum inter-nest distance, Consequently, the central nests are the ones which usually have a high nest density and a small minimum inter-nest distance, whereas the peripheral ones in general have a low nest density and a relatively large minimum inter-nest distance. However, both categories of nests, the 'high density/small minimum internest distance'-ones and the 'low density/large minimum inter-nest distance'-ones, have been shown to be strikingly less successful than nests of medium density and medium minimum inter-nest distance. The relationship of central/peripheral nest position with nest density and minimum inter-nest distance thus easily explains that central and peripheral nests were hardly different with respect to hatching success.

Consequently we conclude that nest density and minimum inter-nest distance are related to hatching success independently from each other; success is highest in nests with the most common distances.

1.3.3.3. *Group size.*

The previous sections have shown that the distribution of nests within a group is related to hatching success. However, the size of a group, *i.e.* the number of nests of which it is composed, might be related to success as well. To check this, groups of different size have been compared in Table 12. Since colonies might differ with respect to success for a

TABLE 12

Hatching success of sandwich terns in groups of different size

Year	Colony	Group	Number of nests	Number of eggs laid	Number of eggs hatched	Hatching success
1970	South	KA	53	97	89	91.8%
		PR	28	44	37	84.1%
1971	West	BO	84	139	93	66.9%
		BE	79	145	103	71.0%
		FM	26	41	30	73.2%

number of reasons, a comparison for success of differently-sized groups can only be made if such groups are taken from the same colony. In 1970 two groups of 28 and 53 nests, respectively, were followed in colony South; both were not significantly different with respect to hatching success (X^2 test: $0.10 < p < 0.20$). In 1971 three groups of 26, 79 and 84 nests, respectively, were followed in colony West. Once more, no significant difference in hatching success was found (X^2 test: $0.70 < p < 0.80$). So we may conclude that within the same colony, within the range of sizes studied, group size is not related in a systematic way to hatching success.

1.3.3.4. *Colony size.*

If colony size is of influence on success, large colonies can be expected to be more successful than smaller ones. Unfortunately, the peak breeders of the population seldom nest in small colonies; for this reason in 1970 and 1971 only relatively large colonies could be compared. The results are given in Table 13a.

In 1971 no significant difference (X^2 test, $0.50 < p < 0.70$) in hatching success was found between colonies of 1150 and 750 breeding pairs, but in 1970 significantly more eggs (X^2 test, $p < 0.005$) hatched in a colony of 550 breeding pairs than in one of 112.

Our data on 1965, 1966 and 1967, summarized in Table 13b, are less complete, but they show a marked difference in hatching success between colonies of over 280, 46-49 and 4-11 nests, respectively. This suggests that colonies of over a few hundred nests are more successful than colonies of the order of a hundred nests or less.

TABLE 13

Hatching success of sandwich terns in colonies of different size

(a)

Year	Colony	Colony size	Nests checked	Number of eggs laid	Number of eggs hatched	Hatching success
1970	NII	112	112	168	125	74.4%
	south	550	81	141	126	89.3%
1971	south	750	61	111	81	73.0%
	west	1150	189	325	226	69.5%

(b)

Year	Colony	Colony size	Nests checked	Number of eggs laid	Number of eggs hatched	Hatching success
1965	5	4	4	4	0	0.0%
	6	4	4	4	0	0.0%
	4	11	11	11	0	0.0%
	3	650	118	158	111	70.2%
1966	B	49	49	51	20	39.2%
	A	1000	692	985	714	72.5%
1967	N 2	9	9	9	0	0.0%
	N 1	9	9	9	0	0.0%
	L K	46	46	58	27	46.6%
	W	280	280	408	301	73.8%

However, the level of hatching success varies between years and a correction for such variations should be made before different years can be compared. Therefore, a hatching success index (HSI) has been calculated for each of the colonies presented in Table 13a and 13b, which has been done as follows: $HSI = \frac{\text{HS of colony} \times 100.}{\text{HS of year}}$ The relationship between HSI and colony size is given in Figure 17. The figure shows clearly that an increase of colony size up to a few hundred nests is correlated with an increase of hatching success. With further increase of colony size, however, no further effect on hatching success is apparent.

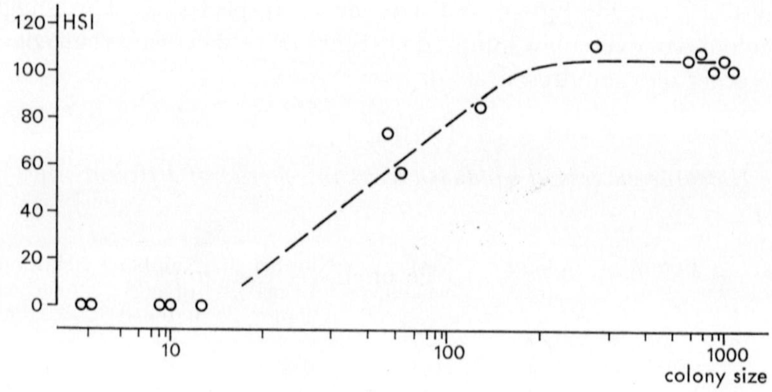

Fig. 17. The relation between colony size (log scale) and hatching success index (HSI). The broken line is drawn by eye. For definition of HSI see text.

Finally, before accepting these results we should verify if variables other than colony size could have been of influence on the relationship between colony size and hatching success. This, however, is only possible for 1970; in 1971 only large colonies were available and the data from 1965, 1966 and 1967 are insufficiently complete to allow a further analysis.

In 1970 the smaller colony N II consisted of only one group of nests. This group was much larger than the two groups studied in colony South (112 *versus* 53 and 28 nests) but in analogy to the preceding analysis an effect on success of this size difference cannot be expected. However, in colony N II group synchronization was stronger than in colony South (see Fig. 16). So, if group synchronization would have caused a difference in hatching success between the two colonies studied, one would expect colony N II to be more successful than colony South. In fact the reverse is true. Therefore, the difference in hatching success between the colonies is more likely to be caused by factors other than group size and group synchronization, such as the crude density of terns over a large area.

Nest density is another factor which might have been of influence on the relationship between colony size and hatching success. To check this, the difference in success between the colonies has been tested for the various classes of nest density. Thereupon the different values of X^2 have been summarized and tested. The results show ($p < 0.005$) that nest density did not cause the difference in hatching success between colony N II and colony South.

1.3.3.5. *Conclusions.*

Summarizing we conclude that group synchronization, minimum inter-nest distance, nest density and colony size are related to hatching success independently of each other. Central *vs* peripheral nest position and group size appeared to have no systematic relationship with hatching success.

A number of aspects of the nest distribution pattern in groups are interrelated: minimum inter-nest distance, nest density and central *vs* peripheral nest position are all positively correlated with each other. Besides, group synchronization is correlated with nest density and central *vs* peripheral nest position: peak breeders usually have a relatively high nest density and often have their nests in the centre of the group.

Considering the strong positive correlation between colony size and hatching success it seems of great advantage for the sandwich tern to nest in large colonies. Within a given colony it seems of no importance whether the terns settle in a large or in a small group, for, within the same colony groups varying in size between 26 and 84 nests were not different in hatching success.

1.3.4. Island synchronization and success.

In the beginning of this chapter the peak and late breeders of the population have been compared (1.3.2); both categories were markedly different with respect to breeding success. Thereupon, to avoid the influence of island synchronization and other season-correlated influences, the analysis of the relationship of the nest distribution pattern in groups and colonies was carried out on nests selected from the population's peak breeders. This analysis showed, that group synchronization, minimum inter-nest distance, nest density and colony size independently of each other were of influence on hatching success. In order to come to a better understanding of the relationship between island synchronization and breeding success, we shall now consider if, and to what extent, variations in nest distribution can explain the difference in success between the peak and late breeders of the population.

Once more it should be emphasized that sandwich tern chicks leave the nest shortly after hatching. Consequently, the distribution pattern of groups and colonies disappears after nest-leaving. This pattern, therefore, can only have influenced the relation between island synchronization and hatching success. A possible relationship of island synchronization with fledging success can be expeected to be independent of this pattern.

Figure 18 gives for 1970 and 1971 island synchronization, clutch size, group synchronization, minimum inter-nest distance, nest density, colony size and hatching success in relation to laying date. Island synchronization is measured as the number of nests per 3-day period. The rate of synchronization of a group has been expressed as the proportion of clutches laid within the laying peak of the group.

The figure shows clearly that the peak breeders of the population had a greater clutch, a higher degree of group synchronization, a smaller minimum inter-nest distance and a higher nest density than the late breeders. They also nested in relatively large colonies. It thus appears that, with respect to hatching success, differences in nest distribution between peak and late breeders are in favour of the first. In the figure the graph representing hatching success shows that the peak breeders indeed hatched a higher proportion of their eggs than the late breeders. However, is it possible to explain these differences in hatching success fully on the basis of the differences in nest distribution? To check this the following analysis has been made: The nests of peak and late breeders, separated as indicated in Figure 18 (the same division has been used in section 1.3.2.), have been arranged according to group synchronization, minimum inter-nest distance, nest density and colony size. Thereupon, for the late breeders a hatching success expectancy has been calculated based upon the hatching success of those peak breeders which had a corresponding nest distribution pattern. Since in both years among peak and late breeders colonies of corresponding size were not available in the population, Figure 12 has been used to account for the influence of colony size on the hatching success expectancy of the late breeders.

The results are presented in Table 14. It shows that in both years the hatching success expectancy of the late breeders is significantly higher than the hatching success observed (X^2 test: 1970— $p < 0.005$, 1971 — $p < 0.005$). Therefore, we may conclude that differences in the nest distribution pattern of peak and late breeders can only partly explain the difference in hatching success. This means that island synchronization is correlated with hatching success apart from group synchronization, minimum inter-nest distance, nest density and colony size.

TABLE 14

Hatching success of peak and late breeders of the island population. 'Expected' = allowing for differences in group synchronization, nest density, minimum inter-nest distances and colony size

1970		Eggs laid	Eggs hatched	Hatching success
Peak breeders		309	251	81.2%
Late breeders	observed	350	122	34.8%
	expected	350	207	59.1%

1971		Eggs laid	Eggs hatched	Hatching success
Peak breeders		436	307	70.4%
Late breeders	observed	253	66	26.1%
	expected	253	120	47.4%

The relationship between island synchronization and fledging success has been studied in 1971 and in part of the 1972 season. To avoid the possible influence of clutch size differences (clutch size strongly decreased in the course of the season) only chicks from one-egg clutches were taken for analysis. Such chicks, originating from various parts of the island, were grouped by laying date. Thereupon chicks from different date groups were compared for fledging success. The results are given in Figure 19.

In both years there was a strong correlation between the number of pairs starting egg-laying within a certain time interval and the subsequent fledging success in that group (Kendall Rank Correlation Test: 1971 — $S = 28$, $n = 9$, $p < 0.005$; 1972 — $S = 62$, $n = 14$, $p < 0.005$).

It is interesting, that in addition the figure shows, that early breeders fledged more chicks than relatively late ones. In both years the graphs representing success decline more sharply than can be explained on the basis of synchronization only. For instance: a comparison of period 1 and 5 in 1971 shows that fledging success was strikingly less for those birds which had laid their eggs in period 5 than in 1, whereas the number of pairs starting egg-laying was about the same in both periods (period 1: 129 pairs, fledging success 72.8%; period 5: 141 pairs, fledging success

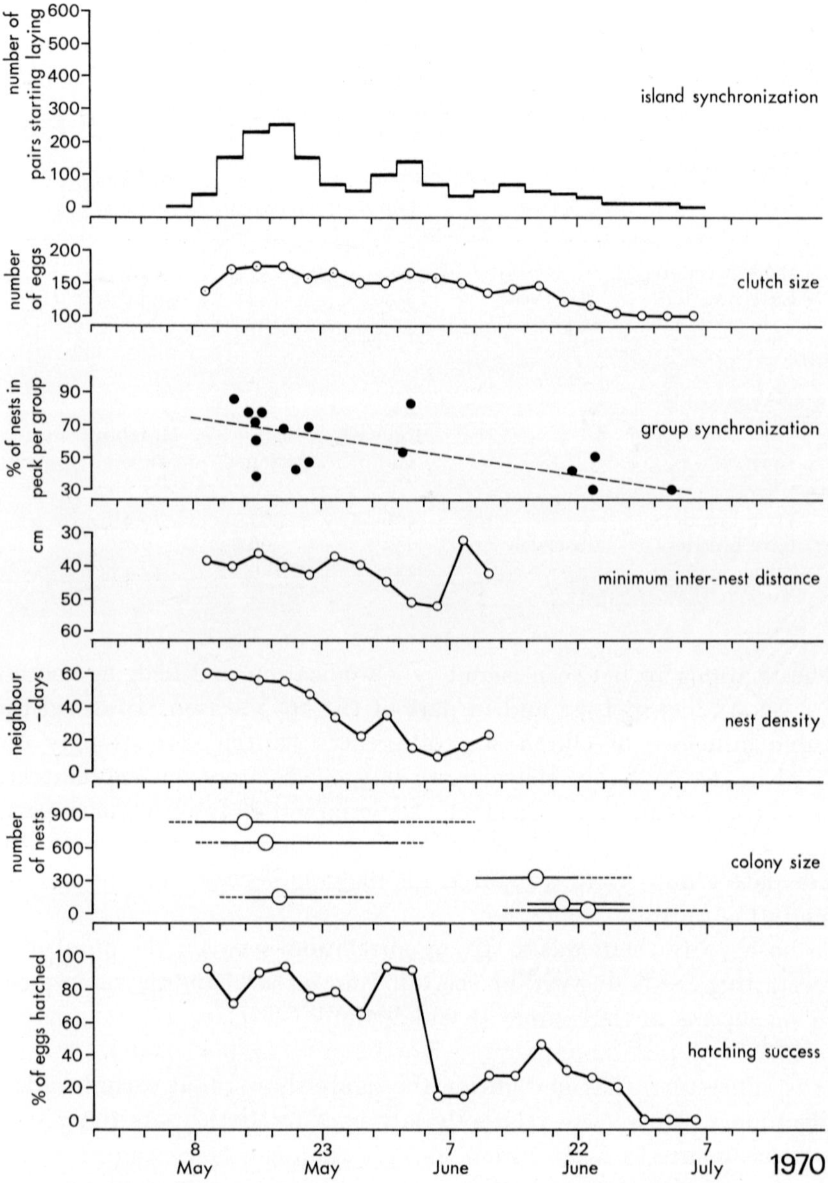

Fig. 18. Characters of nest distribution in relation to laying date. For colony size the open circles indicate the mean, the broken line the full range and the solid line the 80% limits of variation.

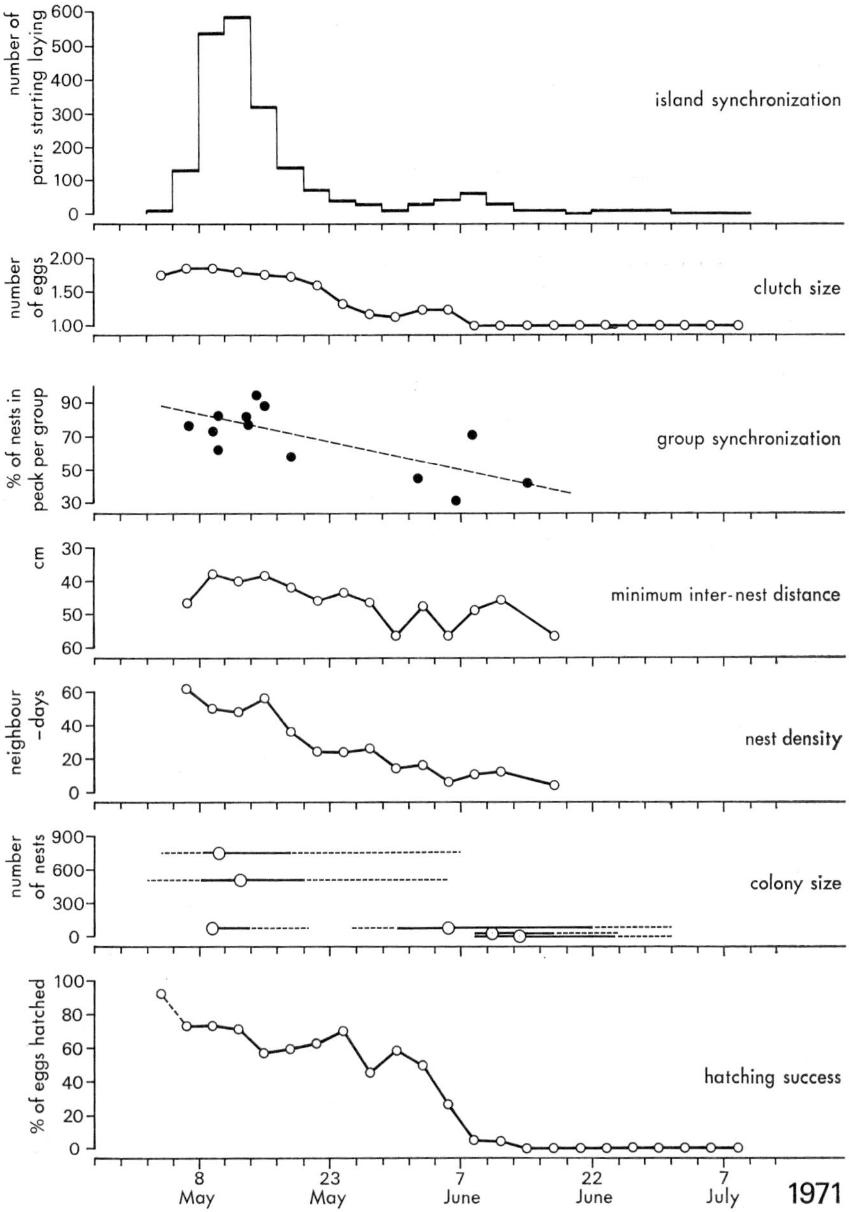
number of pairs starting laying
600
500
400
300
200
100
0
island synchronization
number of eggs
2.00
1.50
1.00
clutch size
% of nests in peak per group
90
70
50
30
group synchronization
cm
30
40
50
60
minimum inter-nest distance
neighbour -days
60
40
20
0
nest density
number of nests
900
600
300
0
colony size
% of eggs hatched
100
80
60
40
20
0
hatching success
8 May
23 May
7 June
22 June
7 July
1971

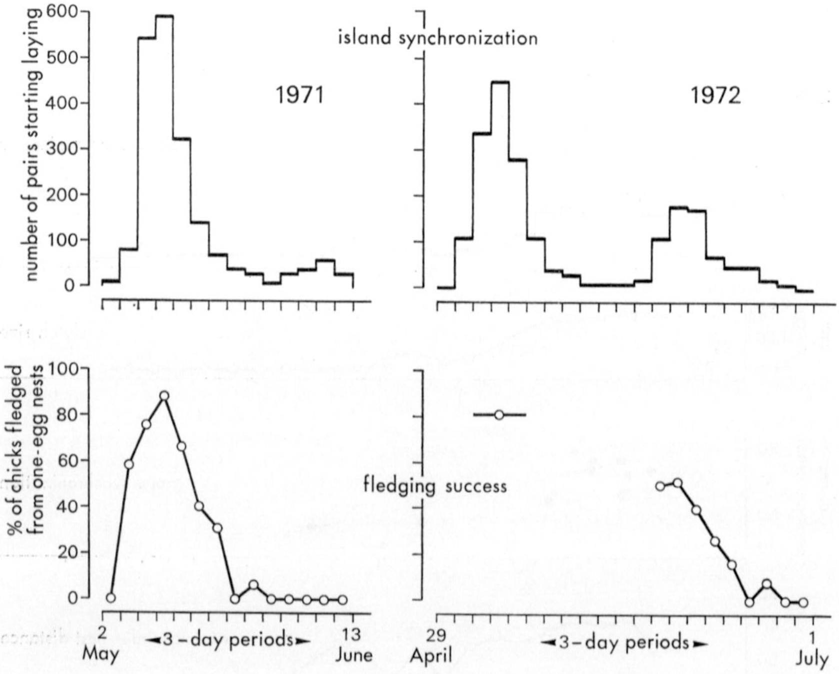

Fig. 19. Relation between island synchronization and fledging success in 1971 and 1972.

52.0%). These results indicate that island synchronization contribute to success in two ways: by promoting simultaneous breeding and by promoting breeding in the most favourable period.

1.3.5. Age and success.

The previous sections have shown, that clutch size and a number of aspects of the nest distribution pattern are of influence on reproductive success. Moreover, these aspects have been shown to be correlated with laying date. As a consequence, the early breeders of the population on the average were much more successful than the late breeders.

The logical question arising from these results is: 'which is the cause of the deviant and unsuccessful behaviour of the late breeders?' In fact, this question leads to a complicated network of aspects of physiology, such as differences in susceptibility to environmental stimuli between individual birds, which is beyond the scope of this study. However, many of such influences can be expected to be related with maturation, which in turn must be related with age. Since, in a number of larid species recruit breeders have been recorded to have a lower breeding success than recidivist breeders (AUSTIN, 1945; COULSON & WHITE, 1958, 1959, 1966), it is important to know if the same holds for the sandwich tern and if so, how birds of different age (on which recruitment depends) are distributed over the population.

Unfortunately, it is not possible to determine the age of adult sandwich terns on external features. In the Griend colonies, however, up to 20% of the breeding birds was found to be ringed, which made it possible to determine the age of this special group. In total 86 ringed breeding birds were caught on the nest; their age varying between 2 and 11 years.

MÜLLER (1959) suggests, that most European sandwich terns do not breed before they are 4 years old. In his migration study, based on 1336 recoveries, younger birds were never recorded from the breeding grounds. We caught many birds of 3 and one of 2 years old, which shows that breeding can start as early as the second year. As only one 2-year old bird was caught it is likely that breeding at this age is exceptional. Therefore, the majority of breeding birds up to and including their third year can be expected to be recruits. It is also likely that the majority of the birds of 5 years and older are recidivists, but 4-year old birds can be either first-time or second-time breeders. This made us decide to distinguish three age classes:

Age class	Age in years	State of breeding	
I	2-3	first time	recruits
II	4	first or second time	recruits and recidivists
III	5+	second time or more	recidivists

In the sandwich tern, both male and female contribute to the care of the brood and it is likely that both will influence breeding success. The number of ringed birds within the population did not allow us to find many pairs in which both mates had a ring. Therefore, we had to correlate breeding success and breeding behaviour of a particular pair with the known age of one partner, either male or female. This procedure may lead to false conclusions, if many pairs are composed of birds of different age classes. However, this seemed not to be the case. On three occasions the age of the members of a pair could be established. The age of the mates in these pairs were 3 + 3, 3 + 4 and 5 + 7 years, respectively.

Table 15 gives the relationship between age and success of 66 birds in 1971. Hatching success increased with age; the differences are statistically significant between class I and class III ($X^2 = 15.09$, df 1, $p < 0.005$) and between class II and clas III ($X^2 = 9.89$, df 1, $p < 0.005$) but not between class I and class II ($X^2 = 1.54$, df 1, $0.20 < p < 0.30$). Fledging success did not differ significantly between the three classes. This might be caused by the small number of chicks hatching in class I and class II and by clutch size differences between the classes. It follows from the table that clutch size increased with age, which probably accounted for the relatively low fledging success among the recidivists.

TABLE 15

The relation between age and reproductive success in 1971

Age class	I	II	III
Age in years	3	4	5-11
Number of nests	12	17	37
Average clutch size	1.33	1.53	1.81
Eggs laid	16	27	67
Eggs hatched	3	10	48
Chicks fledged	2	4	24
Hatching success	18.8%	37.0%	71.6%
Fledging success	66.7%	40.0%	50.0%

In Figure 20, the age is plotted against the date of egg-laying and compared with the distribution of egg-laying in 1971 and 1972. Class III breeders laid their eggs in the highly synchronized peak of egg-laying in the month of May, with only one exception: a 5-year old bird started late in 1971. Class II breeders were found to lay their eggs from early May till the end of June. Class I birds mainly laid their eggs in June and the beginning of July.

Since age is so highly correlated with date of egg-laying, it seems likely that age is also correlated with clutch size and those aspects of nest distribution which have been shown to be correlated with laying date. To demonstrate this, the 3 and 4-year old birds have been taken together and compared with the 5-11 year old ones with respect to clutch size, group synchronization, minimum inter-nest distance, nest density and colony size. The results are summarised in Table 16. It follows that,

TABLE 16

The relation between age, clutch size and nest distribution in 1971

Age in years	3-4	5-11	Early breeders 3-4	Late breeders 3-4
Number of nests	28	34	7	21
Clutch size (% of 2-egg nests)	54%	88%	86%	43%
Group synchronization (% of nests in peak of egg-laying)	36%	76%	71%	24%
Minimum inter-nest distance (% of nests at 21-30 cm)	43%	71%	57%	38%
Nest density (% of nests with 41-120 neighbour-days)	46%	100%	71%	38%
Colony size (% of nests in colonies of over 200 nests)	25%	100%	100%	0%

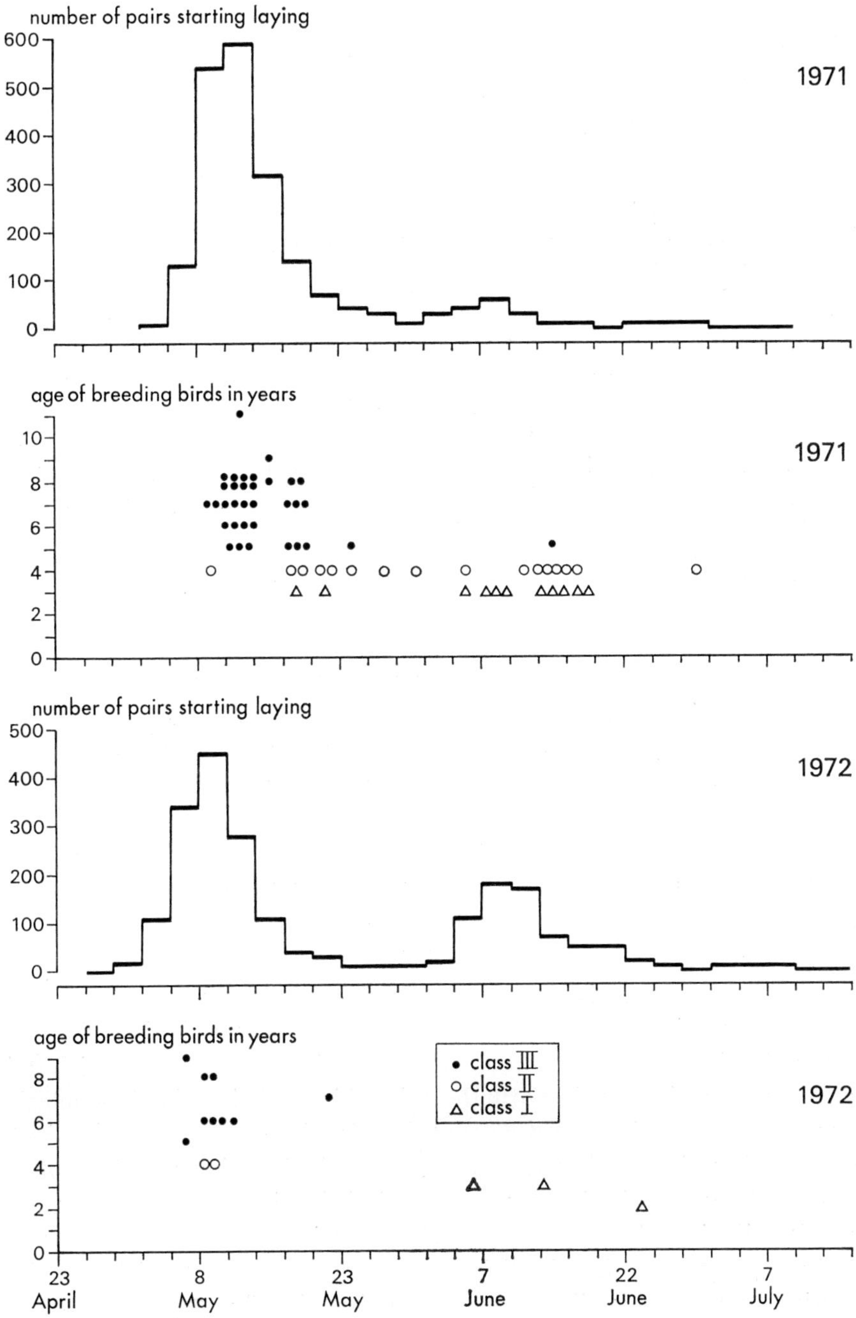

Fig. 20. Temporal distribution of egg-laying compared with age of breeding birds. Griend 1971 and 1972.

with respect to all the characters mentioned, the 5-11 year old birds were indeed markedly different from the 3-4 year old ones. In the right hand of the table the 3-4 year old birds have been separated on the basis of whether they belonged to the peak or late breeders of the population: both groups are markedly different from each other. If we now compare the early breeding 3-4 year old birds and the 5-11 year old ones, it follows that these groups are hardly different with respect to the aspects of nest distribution mentioned in the table. This is of special interest for it shows that the correlation between early laying and nest distribution stands apart from age. However, it has been mentioned before that a small number of 3-year old birds and many 4-year old ones can be expected to have bred before. Since all breeding birds which are known to have bred before, *i.e.* the 5-11 year old group, laid their eggs early in the season, it seems likely that the early breeding 3 and 4-year old birds represent the second-time breeders of their age class. This is supported by the observation of an early breeding 4-year old bird in 1972 which was known to have bred in the preceeding season.

Summarizing, we conclude that sandwich terns which have bred before (3-11 years old) behave markedly different in comparison to first time breeders (2-4 years old). Recidivist breeders lay their eggs early in the season. Moreover, nest aggregation and synchronization of egg-laying are much stronger in the recidivists than in the recruits.

1.4. CONCLUSIONS

Figure 21 presents an overview of the variables found to influence breeding success, with the exception of age and clutch size to be discussed below. The figure shows that sandwich terns nesting at medium nest densities and medium inter-nest distances were strikingly more successful in hatching their eggs than sandwich terns nesting at both higher and lower densities and inter-nest distances. Sandwich terns which laid their eggs synchronously with the other birds of the island population were more successful than the 'asynchronized' breeders. The same holds for breeding birds laying synchronously with the other members of the group. The graphs indicating success of birds with a different degree of synchronization show that date of laying influences success as well; early nesting 'asynchronized' breeders were more successful than late nesting 'asynchronized' breeders. Finally, the figure shows that an increase of colony size up to a certain size (in fact a few hundred nests) is correlated with a marked increase of hatching success. With further increase of colony size, however, no further effect on hatching success is apparent.

The various aspects of nest distribution mentioned above were related to success independently of each other, of clutch size differences, differences in age and, apart from synchronization, also of gross differences in laying date. Central vs peripheral nest position and group size were not related to success independently of the other variables measured; they have not been incorporated in Figure 21.

In addition to the above mentioned relationships we have found that sandwich terns possessing two eggs on the average fledged more chicks than pairs possessing one egg. We also found that recidivist breeders were strikingly more successful than recruit breeders. Differences in success between recruit and recidivist breeders were partly explained by differences in nest density, minimum inter-nest distance, group synchronization, colony size and clutch size. However, other factors such as differences in seasonal correlated environmental influences were likely to have influenced the breeding results of the recidivist and the recruit breeders as well.

1.5. DISCUSSION

With respect to the causation of the differences in breeding behaviour between early and late breeders the results on age are of special interest. It has been shown that age is correlated with first breeding, whereas first breeding is correlated with nest distribution, which in turn is related with success. Not the age of an individual, but whether it has bred before seems to determine breeding behaviour. This suggests that two conditions must be fulfilled before the individual is able to participate successfully in breeding: (a) it must be sexually mature and (b) it must have previous breeding experience.

We have not compared the displays or other sexual behaviour of recruit and recidivist breeders. However, we know that the recidivists arrive earlier in the season than the recruits. Besides, the recidivists are very often mated on arrival, whereas the majority of recruit breeders are unmated when they arrive in the breeding area. In a variety of bird species the onset of migration has been shown to be induced by day-length and the responsiveness to this stimulus seems related with gonadal development. The readiness to respond to social stimuli and to mate during migration might be explained by sexual maturity as well. However, in the kittiwake, a species in which the two members of a pair often mate with one another in successive years, it has been shown that

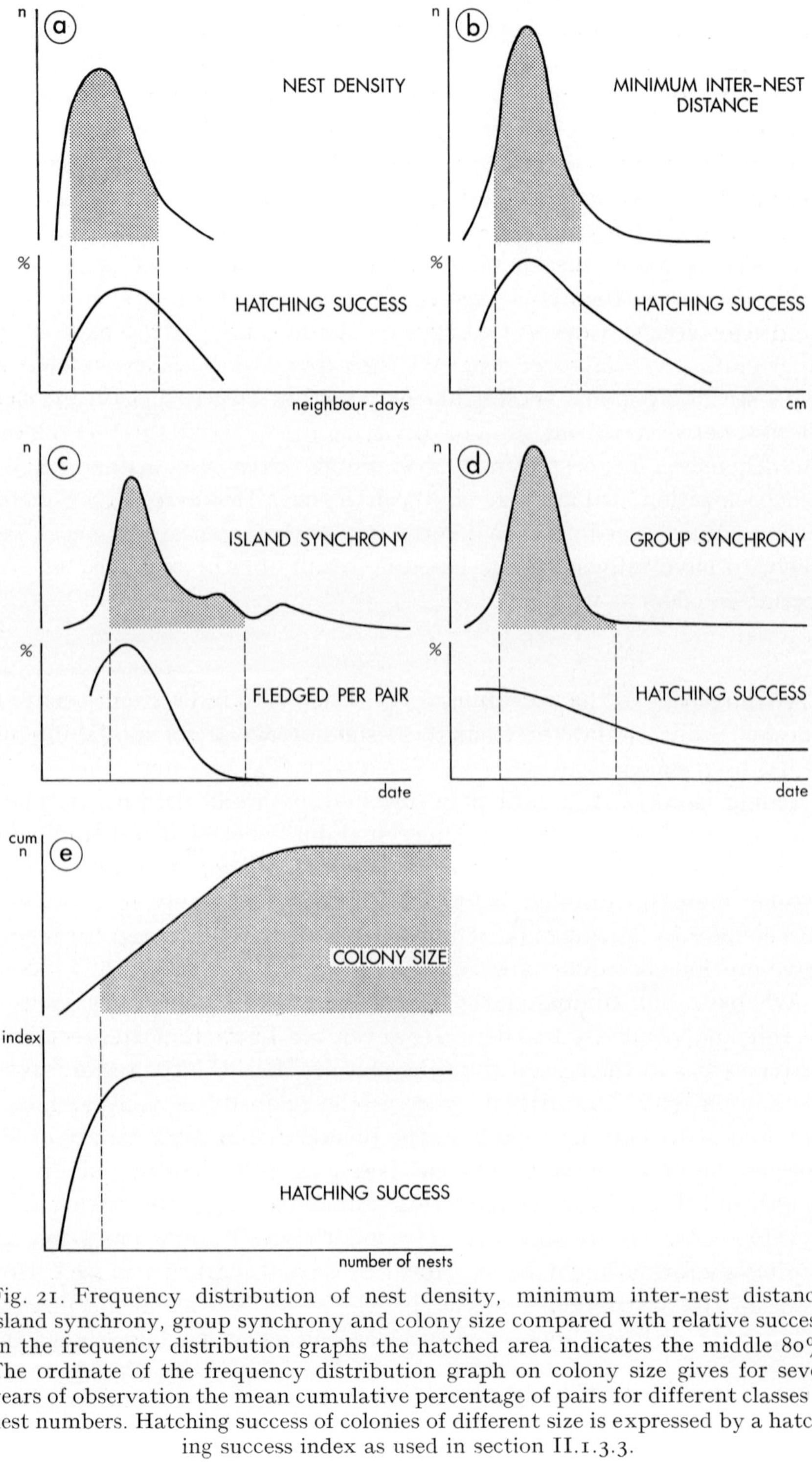

Fig. 21. Frequency distribution of nest density, minimum inter-nest distance, island synchrony, group synchrony and colony size compared with relative success. In the frequency distribution graphs the hatched area indicates the middle 80%. The ordinate of the frequency distribution graph on colony size gives for seven years of observation the mean cumulative percentage of pairs for different classes of nest numbers. Hatching success of colonies of different size is expressed by a hatching success index as used in section II.1.3.3.

egg-laying starts earlier in pairs which have bred together in a previous year than in newly formed pairs of the same age (COULSON, 1966). Obviously, the formation of a pair bond needs more time than its restoration. If the same holds for the sandwich tern, differences in laying date between recruits and recidivists may be caused by differences in experience as well as by differences in sexual maturity.

We have started our analysis with the assumption, that if a particular behaviour pattern has survival value, individuals deviating from this pattern can in general be expected to be less successful breeders. It has been shown that a number of aspects of the spatial and temporal nest distribution pattern were indeed correlated with reproductive success. However, the results obtained cannot prove that these aspects by themselves have survival value; they might be correlated with other aspects, either behavioural or environmental, which are of influence on success. Without going into details, the probability of such relationships will be shortly discussed.

It has been emphasized before that nearly all aspects of the nest distribution pattern are interrelated. In most cases we succeeded to relate the various aspects independently of each other to success. However, the possible influence of colony synchronization on success could not be separated from the relationship between island synchronization and success. Interpreting the results on island synchronization this should be borne in mind.

Analysing the relationship with success of the nest distribution pattern in groups and colonies, the influence of season-correlated environmental influences could be avoided by selecting nests from a relatively small period. Because island synchronization is seasonal, this could not be done in the analysis of the relation between this aspect of nest distribution and success. Although our data (3.4.) strongly suggest that simultaneous breeding and early laying are correlated with success independently from one another, both aspects could not be satisfactorily separated.

Besides seasonal variations, local variations of environmental factors might also have influenced the results. In general such variations — differences in vegetation and in the distance to nesting common terns and black-headed gulls — seemed to play only a minor role in our analysis. However, a slight influence cannot be excluded.

The various aspects of the nest distribution pattern in groups and colonies have been related to hatching success. In fact these aspects can be expected to have had an influence on success till the moment of nest leaving, approximately 4-5 days later. Therefore, they may have influenced the relationship between island synchronization and fledging success. This is not taken into account. However, island synchronization was correlated with hatching succes, apart from the nest distribution pattern in groups and colonies. Thus, it may be assumed that the strong correlation between island synchronization and fledging success was only partly, if at all, caused by variations in the nest distribution pattern of groups and colonies.

Summarizing we may conclude that the analysis of the relationship between island synchronization and succes includes the possible influence of colony synchronization and other aspects of the nest distribution pattern. Besides, island synchronization could not be separated from early egg-laying and, therefore, includes the possible influence of seasonal variations of environmental influences. We have no arguments to accept that the relationship with success of clutch size, group synchronization, minimum inter-nest distance, nest density and colony size were due to incompleteness of the method of analysis.

Finally, it should be emphasized that the present chapter only provides evidence that the features considered typical for the sandwich tern (high density and extreme synchronization in groups, colonies and island population) contribute to success. The data presented give no information which can answer the question 'how' nest distribution contributes to success or, in other words, which requirement imposed by the environment it meets and how it does so. This question can only be answered by studying separately the pressure of different mortality factors.

Data on mortality agents have been collected between 1966 and 1972. The results are presented in Table 17. They have been derived as follows:

Data on egg mortality have been obtained by counting the number of eggs washed away and by checking nests from egg-laying until hatching (II.1.2.). The proportion of eggs classed as 'lost during hatching' refers to eggs disappearing in the hatching period. These eggs were all taken by predators, but a relatively high proportion failed to hatch and was deserted by the terns before they fell victim to predation. Consequently, the primary cause of death remained unknown for most of these eggs. The proportion of chicks preyed upon has been calculated from predation data obtained by direct observation (description of method in chapter I.2). In all years, except for 1972, chicks found dead have been collected and examined on the cause of death. On a few occasions a chick apparently died as a consequence of disease or physical defect. Roughly ten percent of the chicks examined had small wounds on the head and the neck, which were probably caused by the pecks of adult sandwich terns or black-headed gulls. By far the majority of chicks found did not show any external defect, besides that they were all very lean. In 1966 a large number of such chicks was found dead during and immediately after a period of food shortage caused by a long-lasting period of storm. In this case food-shortage was by no doubt the primary cause of death (see I.2.4.). Food shortage might have

TABLE 17

*Data on mortality of eggs and chicks of sandwich terns in the Griend colonies between 1966 and 1972. The figures marked with an asterix are estimates with an accuracy of * 10% and** 15%*

Year	1966	1967	1968	1970	1971	1972
Breeding pairs	1500	1000	1200	1742	2079	2089
Clutch size	1.50	1.65	1.70	1.55	1.75	1.63
Eggs laid	2250	1650	2050	2717	3626	3417
Egg mortality:						
washed away	17.8%	1.8%	0.0%	0.0%	0.0%	0.0%
preyed upon	17.8%	26.1%	10.2%	12.8%	12.7%	6.6%
addled		4.1%	4.8%	5.2%	5.4%	4.3%
embryo died		0.6%	0.4%	0.6%	1.2%	1.0%
egg damaged	11.1%	0.9%	1.2%	2.3%	1.8%	0.4%
egg deserted		0.4%	0.2%	0.1%	1.2%	1.0%
unknown		5.5%	7.4%	9.8%	11.3%	7.3%
Total	46.7%	39.4%	24.2%	30.8%	33.6%	20.6%
Eggs hatched	1200	1000	1557	1860	2408	2713
Chick mortality:						
robbed by black-headed gulls	7.7%	12.0%	6.2%	9.5%	7.8%	5.0%**
robbed by herring gulls	0.0%	4.0%**	2.0%**	9.4%	1.9%	5.0%
found dead	13.7%	4.4%	5.0%	7.1%	15.7%	0.0%
unknown	28.6%	19.6%	29.0%	20.5%	24.8%	29.5%
Total	50.0%	40.0%	42.2%	46.5%	50.2%	39.5%
Chicks fledged	600*	600*	900*	996	1202	1640
	0.40	0.60	0.75	0.57	0.58	0.79

been the primary cause of death of chicks in other years as well. However, ringing results and observations from a hide have shown that the majority of chicks found dead are second-hatched chicks from 2-egg clutches. On the average the second-hatched chick hatches 2-3 days after the first and it is easily deserted by the parents if it is too weak to follow when the nest is left. Very likely most of the chicks found dead were first deserted by the parents and thereupon died of starvation or were killed by the pecks of the adult sandwich terns. The number of chicks found dead was only a part of the total number of chicks that died on the island. Many chicks are likely to have died in dense vegetation after they had left the colony and they will hardly be found. On the basis of the estimated (1966, 1967, 1968) and the calculated (1970, 1971, 1972) total number of chicks fledged within each year, the total number of dead chicks that was not recovered has been calculated. In the table this category has been given as 'unknown'. As it is likely that the chicks taken together as unknown died as a consequence of the same mortality factors as the 'found dead' category, both categories can be regarded as one group. Differences in the chick ratio of the 'found dead' and the 'unknown' category between the different years can be explained by differences in vegetation in the neighbourhood of the colonies and by differences in the effort at which chicks were collected.

The data presented in Table 17 are summarized in Table 18. We may conclude that between 1966 and 1972 mortality of eggs and chicks in the sandwich tern population on Griend was caused by predation (eggs and chicks), high tides (eggs), food shortage (chicks) and a number of factors which might be taken together as hatching failure (eggs), diseases, and physical defects (chicks). Mortality as a consequence of predation, high tides and food shortage may vary considerably from year to year, whereas mortality caused by hatching failure was rather constant.

TABLE 18

Mean proportion of eggs and chicks of sandwich terns lost in the colonies on Griend between 1966 and 1972 (means of percentages given in Table 17)

Mortality agent	Mean proportion of eggs and chicks lost between 1966 and 1972.	
	Eggs	Chicks
Predation	14.3%	11.7%
Washed away	3.8%	
Hatching failure	6.9%	
Deserted	0.6%	
Food shortage, deserted Diseases, physical defects		33.0%

2. PREDATION

2.1. INTRODUCTION

The previous chapter showed that a number of aspects of the nest distribution pattern of the sandwich tern are correlated with reproductive success. Besides, it has been shown that predation was one of the most important mortality factors and it thus seems worthwhile to investigate whether the various aspects of nest distribution analysed have anti-predator significance. However, if nest distribution contributes to reproductive success through lowering predation on eggs and chicks, it can be expected to be part of a complex of anti-predator adaptations, including aspects of habitat selection, attack-escape behaviour, camouflage and many others. For a better understanding of the possible anti-predator function of nest distribution it thus seemed necessary to study thoroughly the species predators and anti-predator behaviour, giving full attention to the effects on predation of associative nesting with common terns and black-headed gulls (see 1.2.3.).

The present chapter gives a survey of the various predators recorded on Griend. For each predator numbers, hunting behaviour and predation figures are given. The influence of each predator species on the breeding results of the sandwich tern in two years of observation is calculated and, finally, the Griend situation is compared with that of other colonies.

2.2. PREDATION BY BLACK-HEADED GULLS

2.2.1. Introduction.

Various authors have mentioned the black-headed gull as a predator on the brood and a pirate on the food of the sandwich tern (Van den Assem, 1954b; Broekhuizen & De Smidt, 1965; Rooth, 1958, 1965).

On Griend the black-headed gull first settled as a breeding bird in 1921. Between 1928 and 1963, in the course of a control program, the wardens on the island collected all the eggs in order to keep the population at a low level. In 1964, 1965 and 1966 only a small proportion of the eggs was collected, the remainder was shaken to avoid continuous re-laying. In 1967 shaking was continued but in each nest a single egg was left undisturbed, which resulted in the fledging of many chicks. In the following years the gulls were left undisturbed as much as possible. Only a few eggs have been taken between 1968 and 1972, mainly for scientific purposes.

Table 1 gives the number of breeding pairs between 1965 and 1972. Immigration from other colonies and the production of chicks on the island can both be held responsible for the steady increase of numbers between 1965 and 1970. Probably a lack of available sites prevented a further increase after 1970.

Apart from the breeding birds, a varying number of non-breeding black-headed gulls use the island as a high tide roost. In April and May their numbers are small, but in July and August up to many thousands may be present. They normally concentrate on the beach and do not visit the higher parts of the island.

2.2.2. Occurrence and behaviour of predators on eggs and chicks.

Van den Assem (1954b) and Rooth (1965) have reported that only a relatively small number of black-headed gulls seemed to have specialized on stealing eggs and chicks of sandwich terns on De Beer (Hoek van Holland). According to Rooth, most of these gulls were in juvenile plumage, which means that they were not involved in breeding activities.

When I first visited the Griend colonies I got the same impression:

a small number of black-headed gulls seemed to keep watch over the tern colonies continually, quickly grabbing an egg or chick (Plate 12 and 13) when an opportunity arose. The behaviour of these predators highly contrasted with the behaviour of other gulls, which apparently did not show any interest in the eggs or chicks of the terns. In contrast with the situation on De Beer, however, the predatory gulls on Griend almost exclusively were in adult plumage.

Three different questions were raised by these observations: (1) what is the actual number of gulls which take part in predation, (2) can we find evidence to distinguish within the gull population specialized predators and non-predatory individuals and (3) can evidence be obtained that predatory black-headed gulls are not involved in breeding activities? These questions were difficult to answer, since the number of observations on actual predation that could be obtained under natural circumstances was very limited. Moreover it was difficult to identify the predatory gulls individually or to follow them for a long period. However, a series of experiments originally set up to study the reactions of the different species of breeding birds on predatory black-headed gulls has provided some interesting information on the behaviour of the predatory gulls involved with respect to the above questions.

Between 7 and 29 May 1966 black-headed gulls' eggs were laid out on the higher parts of the island in a habitat where sandwich terns were expected to start breeding, and in the second half of May on places where they actually did breed. The eggs within these artificial colonies were spaced apart 1-2 m and the area was watched from a hide at a distance which did not cause any disturbance among the birds present. The number of eggs used and the duration of the experiments varied from one case to the other. The artificial set up of the experiments differed greatly from the normal situation in sandwich tern colonies. However, this seemed of minor importance as far as the behaviour of black-headed gulls in response to an exposed egg is concerned. With respect to the eggs used, it should be noted that previous experiments (BROEKHUIZEN & DE SMIDT, 1964; VEEN, unpubl.) had shown that black-headed gulls do not prefer either sandwich tern eggs or eggs of their own species, if both are presented simultaneously. Both eggs might be different with respect to camouflage, but since our data were only used in a comparative way, this would not interfere with our conclusions.

The results are summarized in Table 19. It is shown, that the number of black-headed gulls involved in predation increased in the course

Plate 6. Sandwich terns threatening one another. The picture clearly illustrates that neighbouring birds nest literally within pecking distance of each other.

Plate 7. Sandwich tern nests conspicuously revealed by white faeces.
(phot. J. van de Kam).

6

7

TABLE 19

Predation by black-headed gulls on eggs laid out during experiments in 1966

Date	Species of bird* present	Duration of observation in hours	Number of eggs laid out	Predation** latency	Number of gulls preying upon the intact eggs
7/5	*C B* O	3,5	5	—	0
8/5	B O	5	5	—	0
9/5	C B *O*	3,5	5	—	0
12/5	*A* B O	7,5	5	—	0
17/5	*C B* O	3	3	20 (20)	1
21/5	*B S*	5	5	40 (40)	2
26/5	*C B S*	5	8	2 (2)	4
27/5	*B S*	5	8	5 (20)	> 5
29/5	*C* B *S*	4	6	6 (20)	> 3
29/5	*B S*	2	5	4 (29)	> 4

* The capitals refer to Arctic tern (A), Common tern (C), Black-headed gull (B). Sandwich tern (S), and Oystercatcher (O) present in the area of the experiment. Italicized capitals mean that the birds are incubating.

** Predation latency = Time in minutes between start of experiment and first predation attempt and (between brackets) first egg broken.

of May, while the time lapse between the start of the observation and the first predation attempt became increasingly shorter.

Before interpreting these results, a short description of the behaviour of the predatory gulls is needed:

Between 7/5 and 12/5, in the beginning of the tests, the gulls predominantly reacted with fear to the eggs. The eggs were either avoided or carefully approached and thereupon inspected from all sides. After some time, interest and fear gradually decreased and the gulls more and more ignored the eggs, even when standing, walking, sleeping and preening at short distance from them. On 7, 8 and 9/5 a few eggs were taken by oystercatchers. On 7/5 the gulls completely ignored the remains of the eggs broken by the oystercatchers. On 8/5 two gulls were pecking at an empty egg-shell and on 9/5 several gulls quarreled with the oystercatchers over the contents of the eggs. None of the gulls, however, touched the intact eggs until 17/5.

On 17/5 an adult black-headed gull which hesitated for several minutes, finally pecked a little hole in two different eggs. From one of these eggs part of the contents was eaten. Later on several gulls (at least 6) ate from the remains, threatening and fighting one another.

On 21/5 the behaviour of some gulls had markedly changed; 40 minutes

Plate 8. A sandwich tern on the nest with its two chicks. (phot. J. van de Kam).

after the start of the experiment an adult gull suddenly alighted, immediately broke an egg and ate from the contents. Two other eggs were taken in the same way.

The results of tests carried out after 21/5 all have in common that several gulls still ignored the eggs, whereas others immediately tried to rob them. Of special interest is a test made on 26/5 near a sandwich tern colony heavily preyed upon. At the start of this test 4 black-headed gulls were present alongside this colony and within 10 minutes all 'experimental eggs' had been taken by these gulls.

The above observations strongly suggest that individual black-headed gulls develop into experienced egg-predators through a quick learning process. Probably most of these gulls get acquainted with an egg as a food source when it is already broken. In our experiment oystercatchers might have started the process by breaking the first eggs. Thereupon a small number of predatory black-headed gulls might have effectuated that many other gulls became aware of the eggs as food source. According to the observations on 26/5 experienced predators keep an eye on places where they have been successful before.

In 1966 the number of predatory black-headed gulls was checked throughout the whole season. An increase in numbers as recorded in Table 19 for the month of May continued over the month of June till an estimated maximum of approx. 20 'experienced egg-predators' was recorded in the middle of July. In the second half of July and in August, when the majority of sandwich tern eggs had hatched, numbers gradually decreased. Comparable observations were made in other years of observation, the number of egg-predators ranging from only a few individuals in the beginning and the end of the season up to a maximum of a few tens in the middle of the season.

With respect to the number of black-headed gulls preying upon chicks, some interesting information was obtained in 1970. In May of that year I took all eggs from 15 nests of black-headed gulls. Observations from a hide showed that the majority of gulls concerned re-laid. A few pairs, however, gave up and settled in the neighbourhood of a sandwich tern colony where they established new territories. At the end of the breeding season in three of such territories concentrations of pellets were found containing the rings of 93 sandwich tern chicks. Thereupon all suitable places on the island were searched for pellets of black-headed gulls with chick-remains. Altogether only 9 rings were found spread over the whole area. Some rings may have escaped our attention. However, as many rings were red coloured and therefore very conspicuous, it is not likely that concentrations of pellets with a great number of rings, as found in the above described situation, have been overlooked. We therefore conclude that only three or a few more black-headed gulls were actively involved in predation on chicks in 1970.

Unfortunately we have no reliable figures on the exact number of chick-predators in other years of observation. Their numbers, however, always remained far below the maximum number of egg-predators.

In all years of observation predatory black-headed gulls were never seen with a nest of their own. Moreover, the remains of eggs and chicks in pellets of black-headed gulls were always found on gathering places of non-incubating black-headed gulls. In 1970 we found that predation on chicks was practized by a few gulls which had failed to re-lay after having lost their clutch early in the season. It follows from this information, that predation on eggs and chicks of sandwich terns is mainly done by black-headed gulls which themselves are not involved in breeding activities.

2.2.3. Predation on eggs.

Methods.

In all years of observation groups of nests have been selected for detailed study. From egg-laying until hatching the nests of these groups were checked and their contents noted (for details on methods see II.1.2.). Eggs disappearing in the incubation period were classified as robbed. Although there was no doubt whether an egg had been taken by predators or not (remains of egg-shells or contents of the eggs were nearly always present), our method failed in giving information on the species of predator involved. However, black-headed gulls and oystercatchers were the only egg-predators active on the island. Data obtained by observations from a hide showed that the oystercatchers contributed to egg-predation only on a small scale (always less than 10%) and our predation figures, therefore, can be regarded closely approaching the actual number of eggs taken by black-headed gulls.

Up to 1967 the nests were checked daily, but in 1967 we discovered that human disturbance sometimes influenced predation: black-headed gulls and oystercatchers robbed eggs during our visits. To avoid this unpleasant influence as much as possible, disturbance was minimized in the following years by checking nests every three days and by speeding up the proceedings. Observations showed that with this improvement predation was no longer increased because of the nest check.

Thus, in 1966 and 1967 human disturbance might have influenced predation by exposing eggs and chicks to predatory black-headed gulls. Moreover, human disturbance might have increased the number of predatory gulls in the same years by collecting the eggs of the gulls (see II.2.2.2.) Since it was impossible to measure the effects of human disturbance on the predation figures in 1966 and 1967, the results from both years have to be considered carefully. Unless otherwise mentioned, they have been excluded from the results given below.

Results.

Table 20 gives the predation results of groups of nests checked in 1970 and 1971. It is shown, that the proportion of eggs taken by black-headed gulls (strictly speaking the proportion which disappeared from the nest; see methods) differed greatly between the different groups within a certain year. Since these differences seemed to be mainly related to the mean date of egg-laying, the proportion of eggs taken by the gulls

TABLE 20

Egg predation in colonies of the sandwich tern in 1970 and 1971

1970

Total number of breeding pairs	Colony	Nests checked	Eggs laid	Eggs preyed upon	Mean date of egg-laying
550	South (group 1)	28	46	2 (4.3%)	13/5
	(group 2)	64	112	0 (0.0%)	16/5
250	North point	38	73	2 (2.7%)	16/5
154	North IIR	154	212	19 (9.0%)	22/5
26	West IIR	26	31	14 (45.2%)	21/6
15	South point	15	18	16 (88.9%)	24/6
15	West IA	15	15	14 (93.3%)	4/7
	Total	340	507	67 (13.2%)	

1971

Total number of breeding pairs	Colony	Nests checked	Eggs laid	Eggs preyed upon	Mean date of egg-laying
750	South	68	122	8 (6.6%)	11/5
1150	West (group 1)	94	165	19 (11.5%)	14/5
	(group 2)	25	37	7 (18.9%)	17/5
	(group 3)	26	34	15 (44.1%)	2/6
32	North point II	40	53	20 (37.7%)	2/6
	Total	253	411	69 (16.8%)	

is given in relation to the laying date in Figure 22. The figure also gives the distribution of egg-laying of the island population. It follows that in both years the percentage of eggs taken by black-headed gulls strongly increased in the course of the laying period.

In Figure 22 the predation percentages are given for 12-day periods. For each period, the nests that were kept under observation and on which each percentage is calculated, could be regarded as a representative sample of the island population. The various predation percentages, therefore, can be extrapolated to the island population. For both years the number of eggs robbed over the total number laid has been calculated for each period. The ratio of the sum of the number of eggs

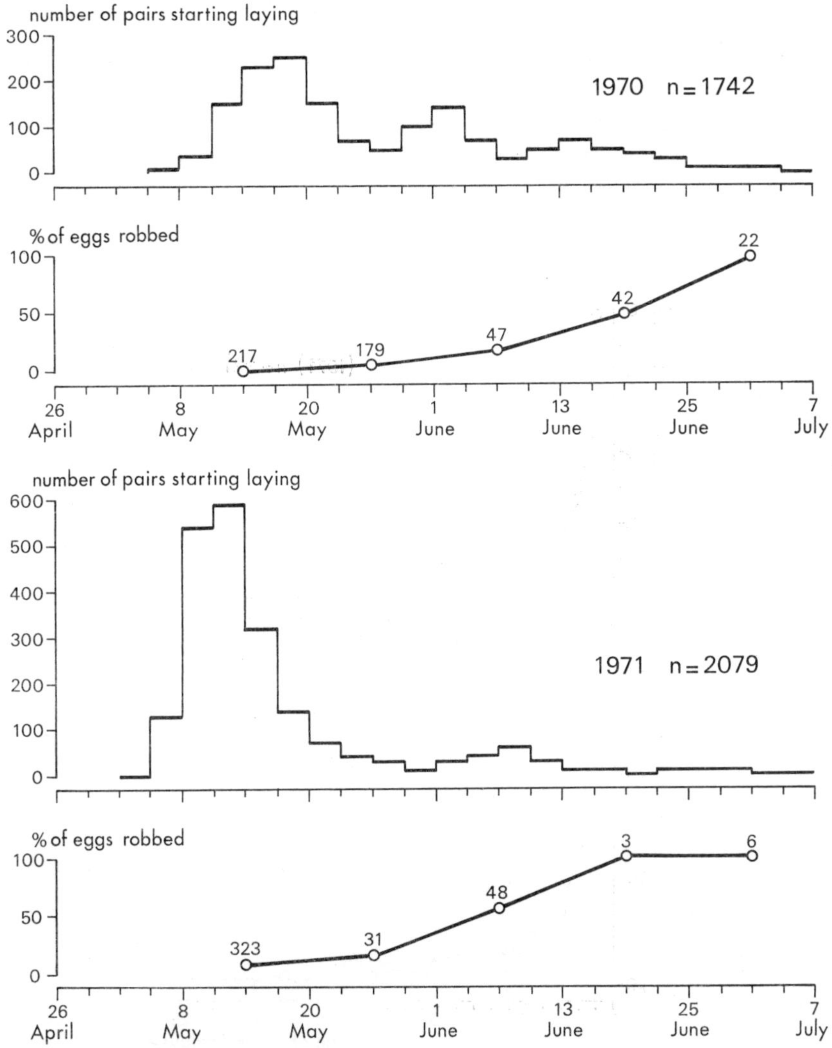

Fig. 22. Distribution of egg-laying of sandwich terns in 1970 and 1971 and the proportion of eggs robbed by black-headed gulls. The number of eggs on which each percentage is calculated is given with each point.

robbed for all periods of a certain year and the total number of eggs laid within that year gives an impression of the 'overall predation percentage' of the year, viz. 348/2717 = 12.8% in 1970 and 460/3626 = 12.7% in 1971.

Figure 23 gives the relation between the percentage of eggs robbed from nests followed in 1967 and 1970 and the age of the clutch in days. In both years the majority of eggs was taken during the first days after laying. (The proportion of eggs robbed in 1967 is much higher than in 1970, which might be partly due to human disturbance.) To interpret these results it is necessary to give some detailed information on the circumstances in which the eggs were taken by the gulls. Altogether I witnessed more than a hundred times black-headed gulls taking an

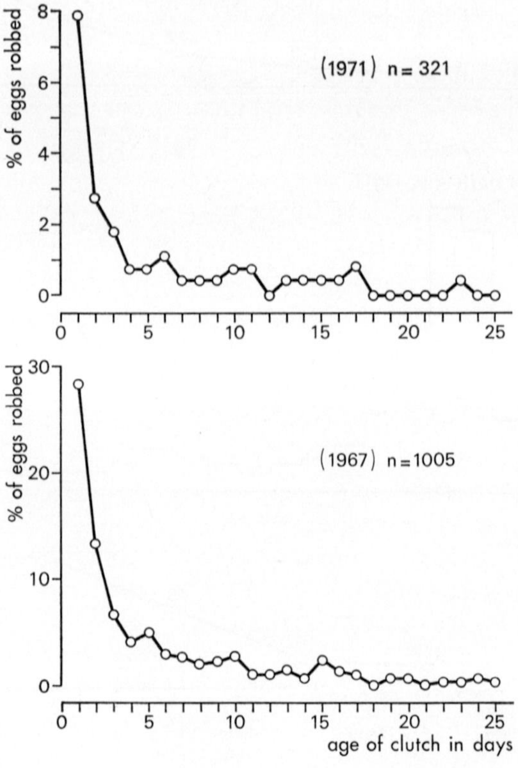

Fig. 23. Relation between the percentage of eggs robbed and the age of the clutch in days.

egg from the sandwich tern colonies. In all occasions this was done in the absence of the nest owners. The circumstances in which it occurred may be summarized as follows:

(1) Nests left voluntary: During the first days of incubation, when the clutch is not yet completed, the nest is often left for courtship display and copulation.

(2) Nests left during disturbances: Throughout the season the terns incidentally fly up from the nest in response to a variety of disturbing circumstances. The tendency to fly up, however, is relatively strong during the first days of incubation.

(3) Deserted clutches: Many cases of egg-stealing refer to a situation in which the clutch was completely deserted. During the first days after egg-laying the clutch is sometimes deserted as a consequence of serious disturbance. At the end of the incubation period the clutch is deserted when the eggs fail to hatch, or, in two-egg clutches, when the parents leave with the first-hatched chick. Breeding birds which have started egg-laying late in the season often desert their clutch at any time during incubation. Sometimes, because the nest is overgrown with high vegetation. Usually, however, the nest is left for unknown reasons.

The circumstances in which the eggs were robbed can easily explain the high percentage of eggs taken during the first days of incubation. As a consequence of early leaving in two-egg clutches one would expect an increase of the percentage of robbed eggs once more at the end of the incubation period. Such an increase exists after the 25th day of incubation. After the 25th day of incubation, however, predation on eggs could not be measured accurately and therefore the results presented in Figure 23 have only been given up to and including the 25th day.

For a judgement of the effect of predation on breeding success we shall need an assessment of the viability of the eggs taken by the gulls in the course of the season. Considering the circumstances in which predation was observed, the eggs taken at the end of the incubation period and the eggs taken from breeding birds which started laying at the end of the season are likely to have been deserted by the terns. Therefore, if not taken by the gulls both categories of eggs would have died from other mortality factors and with respect to the influence of predation on breeding success they can be regarded as non-viable. In 1970 and 1971 the proportion of eggs of both categories within our data on predation have been extrapolated over the population giving full account of seasonal changes in predation as given in Figure 22. Table 21

TABLE 21

Predation by black-headed gulls on different categories of eggs of the sandwich tern

	1970	1971
Number of eggs laid	2717	3626
Eggs robbed at the end of incubation	106 = 3,9%	199 = 5,5%
Eggs robbed from late breeders (egg-laying after 13 June)	168 = 6,2%	66 = 1,8%
Others eggs robbed	74 = 2,7%	195 = 5,4%
Total robbed	348 = 12,8%	460 = 12,7%

gives the results. It follows, that only 2,7% (1970) and 5,4% (1971) of viable eggs were taken from the population, whereas 10,1% (1970) and 7,3% (1971) of non-viable eggs were taken.

2.2.4. Predation on chicks.

Methods.

A few days after hatching the sandwich tern chick leaves the nest. This behaviour made it impossible to assess chick predation in a way as had been done in the case of egg predation, for any chick might disappear from its nest as a result of either predation or leaving. For this reason, data on chick predation had to be gathered by direct observation.

In 1966, 1967, 1968, 1970 and 1971 many hours were spent in a hide, watching a number of nests with chicks of known age. The observations provided some interesting details on the situation in which chicks were robbed and allowed an estimate of the proportion of chicks taken by the gulls.

In addition to the results obtained by direct observations, in 1970 ring recoveries from black-headed gull pellets have given some interesting information.

Results.

Altogether I have at my disposal 70 observations in which black-headed gulls robbed a chick from the sandwich tern colonies. In all occasions the chicks taken by the gulls varied in age between 0 and 3 days.

Between 1966 and 1971 chicks of known age have been watched from a hide on different days and in different colonies. Predatory black-headed gulls appeared to be active from 4.00 a.m. to 21.00 p.m., which means that each chick up to the age of three days was exposed to predation for 51 hours. Based upon these data, the number of chicks taken by the gulls within a certain time-span allowed a calculation of the percentage of chicks taken up to the age of 3 days. Since all observations of a certain year had to be taken together, because the number of chicks preyed upon was always small, the method had the disadvantage that possible differences in chick-predation between the colonies were not taken into account, as were possible differences in vulnerability related with age. However, differences in predation on chicks between the colonies can be expected to be of minor importance, since predatory black-headed gulls were known to roam the whole island, frequently visiting each concentration of chicks. Differences in vulnerability related with age are likely to occur, but usually all age classes were evenly distributed within our data.

In 1967 predation seemed to be relatively high shortly after I had entered the hide. This phenomenon might be related to gulls having specialized on predation during human disturbance, which had been noticed in the case of egg-predation in the same year (see II.2.2.2.). To avoid this influence within our data the first 30 minutes of each observation period have been excluded from the calculation in 1967.

Finally, human disturbance might have accelerated leaving, which in turn might have increased predation. The observations, however, were done from a hide, which caused negligible disturbance, and this factor is not likely to have been of influence in any important way.

The results are given in Table 22. Between the different years predation varied from 6,2 to 12,0%. Whether the figures represent the true situation in each year is difficult to decide within the unknown limits of accuracy of our method. However, the high percentage in 1967 coin-

TABLE 22

Proportion of sandwich tern chicks taken by black-headed gulls in different years of observation

Year	Number of observations in chick-hours	Number of chicks robbed	Percentage of chicks robbed*
1966	12516	19	7.7%
1967	2117	5	12.0%
1968	2475	3	6.2%
1970	1076	2	9.5%
1971	2632	4	7.8%

* The percentage of chicks robbed has been calculated from the number of chick-days and the number of chicks robbed, assuming that each chick is exposed to predation for 51 hours (see text).

cided with an exceptionally high predation on eggs in the same year. In 1966 our results can be compared with direct observations on 48 chicks watched each day from 4.00 to 21.00 hours between hatching and leaving. In this case 4 out of 48 = 8,3% was robbed, which is well in accordance with the 7,7% of the table. Based on these facts, the results given in the table seem to be rather a good estimate of the actual predation figure in each year.

Table 23 gives the results of ring-recoveries obtained from pellets of black-headed gulls and from chicks found dead in the colonies at an age of 0-3 days (period of ringing 7/6-7/7 1970). From the pellets of the gulls it follows that second-hatched chicks of a clutch of two were more heavily preyed upon than first-hatched chicks (X^2 test: $p < 0.005$). The differences between first-hatched chicks of two-egg clutches and chicks of one-egg clutches were not significant.

The data on chicks found dead in the colonies follow the same line as those obtained from the pellets of the gulls, *i.e.* second-hatched chicks were found much more often than first-hatched chicks or chicks from one-egg clutches (X^2 test: $p < 0.005$); both latter categories not being statistically different.

TABLE 23

Ring recoveries obtained from black-headed gull pellets and from chicks found dead in the colonies on the age of 0-3 days

Clutch size	Category of chick	Number of chicks ringed	Number of rings recovered from: Black-headed gull pellets	Chicks found dead in colony
1 egg		162	3 (1.9%)	2 (1.2%)
2 eggs	first-hatched	792	11 (1.4%)	4 (0.5%)
2 eggs	second-hatched	495	57 (11.5%)	50 (10.1%)

For the interpretation of these results it is necessary to give some detailed information on the circumstances in which chicks are taken by the gulls. In 70 observations only once a chick was taken from the nest when both parents were present. In all other occasions they were taken when the nest was either temporarily left or deserted, which mainly occurred in the following situations:

(1) Incidentally sandwich terns fly up from the nest in response to

disturbing circumstances. 15 out of the 70 observations in which chicks were seen to be taken by black-headed gulls refer to this situation.

(2) By far the majority of chicks which fall victim to the black-headed gulls are taken at the moment of leaving the colony or shortly afterwards. Leaving occurs 3-6 days after hatching of the first egg. Then the first-hatched chick is, because of its size, out of danger to be taken by the gulls. Within a clutch of two, however, the second-hatched chick usually hatches 2-3 days after the first. It is therefore at the moment of leaving still relatively small and weak. It often straggles behind and is then taken by the gulls. Sometimes the moment of leaving the colony is accelerated under the pressure of disturbance and then chicks may leave as much as two days earlier. In such a situation the risk to be taken by black-headed gulls highly increases for both first- and second-hatched chicks, but especially for the latter.

The fact that so many chicks are robbed by black-headed gulls at the moment of nest-leaving easily explains the high proportion of second-hatched chicks in the pellets of these gulls. Second-hatched chicks occurred in equal percentages in the pellets of the gulls and within the category found dead. This suggests that different mortality factors might have affected the chicks in the same way. However, observations from the hide have shown that the majority of chicks found dead in the colonies failed to leave the nest together with the parents and first-hatched chick; they were deserted and thereupon died of starvation. We know that the majority of chicks preyed upon by black-headed gulls belonged to the same group of stragglers. This means that the majority of chicks which was taken by these gulls would have otherwise died of starvation.

Table 24 gives the 1970 ring-recoveries, arranged with respect to the date of ringing (= hatching). For first-hatched chicks as well as for second-hatched chicks, the results point in the same direction: In the course of the season, the proportion of rings recovered from pellets of black-headed gulls increased, whereas the proportion of rings recovered from chicks found dead decreased.

An increase of the proportion of chicks taken by black-headed gulls could be expected as a result of specialisation of the predators, in combination with a decrease in the number of chicks available. If we realise that the chicks preyed upon and found dead mainly belonged to the same group of deserted individuals, an increase of the pressure of predation logically effects a decrease of the proportion of chicks found dead.

The proportion of rings recovered from the pellets of the gulls be-

TABLE 24

Ring recoveries obtained from black-headed gull pellets and from chicks found dead in the colonies (age 0-3 days) at different times of the season in 1970

First-hatched chicks

Ringing period	Number of chicks ringed	Rings-recovered from black-headed gull pellets	Rings recovered from chicks found dead
2/6-13/6	586	6 (1.0%)	4 (0.7%)
14/6-25/6	281	1 (0.4%)	1 (0.4%)
26/6- 7/7	198	7 (3.5%)	1 (0.5%)
8/7-18/7	69	7 (10.1%)	0 (0.0%)

Second-hatched chicks

Ringing period	Number of chicks ringed	Rings recovered from black-headed gull pellets	Rings recovered from chicks found dead
2/6-13/6	180	16 (8.9%)	32 (17.8%)
14/6-25/6	111	7 (6.3%)	7 (6.3%)
26/6- 7/7	216	34 (15.7%)	12 (5.6%)
8/7-18/7	8	0 (0.0%)	0 (0.0%)

tween 14/6 and 25/6 is relatively low in view of the general trend of our data. Although the supporting data are few, this might be a point of interest. Between 14/6 and 25/6 ringing was carried out on a much smaller scale than in the preceding and following period and we can expect that human disturbance was relatively less between 14/6 and 25/6. Therefore, a relatively low figure of predation within the given period suggests an influence of human disturbance on our data, *i.e.* flying up of the adults or early nest-leaving of the chicks caused by my visits to the colonies.

For a judgement of the effect of predation upon breeding success we shall need an assessment of the viability of the chicks taken by the gulls in the course of the season. In 1970 9,5% of the chicks were taken from the colonies (Table 22). It follows from the results of pellet analysis (Table 23) that 57/71 (= 80%) of the chicks robbed were second-hatched chicks from two-egg clutches. These chicks can be expected to have died from starvation if they had not been taken by the gulls.

Thus, in 1970 black-headed gulls robbed approximately $57/71 \times 9{,}5\% = 7{,}6\%$ non-viable and 1,9% viable chicks from the population. In 1971 7,8% of the chicks was taken from the colonies. No rings were recovered from the pellets of black-headed gulls in this year and the proportion of first and second-born chicks found dead in the colonies has been used to assess the ratio of non-viable and viable chicks within the proportion preyed upon, viz.: 6,3% and 1,5% from the total population.

Summarizing we conclude that by far the majority of chicks robbed by black-headed gulls would have died by other mortality factors if they had not been taken by this predator.

2.2.5. Occurrence and behaviour of food pirates.

It has been mentioned before, that predation by black-headed gulls on eggs and chicks of sandwich terns was restricted to a relatively small number of experienced predators. With respect to the food of the terns, however, we got the impression that any gull tended to be a pirate if circumstances were favourable. The attractiveness of the food of the terns was clearly demonstrated by observations in which breeding gulls were seen to fly up from their nest to steal a fish from a nearby tern and feeding it to their own chicks afterwards. In general, however, black-headed gulls involved in breeding activities took part in piracy on a very small scale and fish-stealing developed into a habit in non-breeding individuals only.

In all years of observation fish-pirates could be classified in two more or less distinct groups: the 'specialists' and the 'occasional pirates'. The specialists were small in numbers. Usually up to 10 or 20 individuals were present each year. They kept watch over the colonies from early morning till late in the evening and probably lived completely on the fish supplied by the terns. The occasional pirates highly outnumbered the specialists. When sandwich tern chicks were abundant, more than a hundred might be actively involved in piracy at the same time. Fish-stealing, however, was only a part-time business for these gulls. They only visited the tern colonies when their main feeding grounds on the tidal flats were out of reach, *i.e.* at high tide and in periods of high water following stormy weather (see I.2.4.2.).

Figure 24 gives the relation between the number of fish-pirates in colony West 1967 and the tidal movements. It follows, that the number of gulls present varied greatly as a result of the daily rhythm of the occasional pirates.

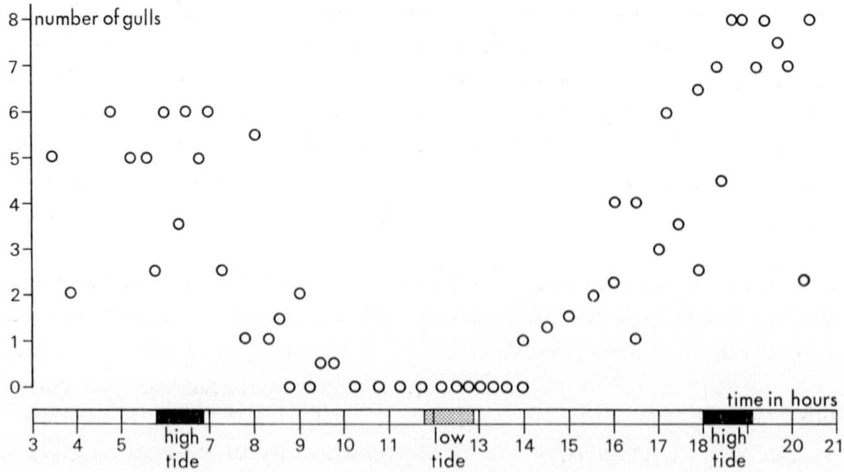

Fig. 24. Numbers of fish-stealing black-headed gulls present in part of the sandwich tern colony in relation with the tide (colony West, 18-19 June 1967).

2.2.6. Intensity of piracy.

Methods.

In order to assess the proportion of fish taken by black-headed gulls, observations were made from a hide in different parts of the colony in 1966, 1967, 1968 and 1970.

In 1966 groups of nests with chicks of known age were selected for detailed study. On arrival, fish-carrying adults were identified in relation to their nest and thereupon followed until disappearance of the fish. For each fish, the final destiny, the species and the estimated length were recorded. (The method of estimating length and calculating weight has been given in I.2.4.2.)

In 1967, 1968 and 1970 fish-carrying terns have been followed without paying attention to individual nests. In addition to the number of fish robbed, data were collected on the occurrence, behaviour and predation success of black-headed gulls involved in stealing fish.

Results.

In all years of observation piracy on fish (Plate 14 and 15) varied from place to place in relation with the distribution of chicks. In general piracy was high as long as the chicks were concentrated in the colony. Piracy decreased when chicks left the colony and fish pirates almost completely ignored the food provided for chicks which had spread over the island. On many occasions piracy seemed to revive in the end of the season, when fledgelings concentrated along the beach at high tide.

In 1966, 1967, 1968 and 1970 fish-carrying terns have been followed as described under methods. In all years observations were done on fish intended for chicks present at the nest, consequently our data can be expected to refer to a situation in which the chicks' supply of food was maximally influenced by black-headed gull predation. The observations were made on different times of the day, giving full account of differences in piracy caused by a change of the number of pirates in relation with the tide. The results are shown in Table 25. The category 'various destinies' refers to fish refused by the chicks, robbed by common and arctic terns (less than 1.5% in all years) or robbed by sandwich terns (4.7%, 0.0%, 0.5% and 8.3% from 1966 to 1970). The table shows that the proportion of fish taken by black-headed gulls gradually increased from 17.8% in 1966 to 37.5% in 1970. This increase of piracy coincided with a relative growth of the black-headed gull population, expressed as the black-headed gull/sandwich tern ratio in Table 25. We cannot escape the impression that the correlation between the gull/tern ratio and the proportion of fish stolen was at least partly causal, since it is likely that the number of occasional pirates was directly related with the size of the gull population.

TABLE 25

Proportion of fish taken by black-headed gulls in the sandwich tern colonies in different years of observation

Period of observation	1966 (10/6-3/7)	1967 (7/6-20/6)	1968 (30/5-8/6)	1970 (14/6-21/6)
Number of fish	2524	1552	1026	336
Fed to chicks	75.7%	75.2%	60.1%	52.1%
Robbed by black-headed gulls	17.8%	23.5%	34.8%	37.5%
Various destinies	6.5%	1.3%	5.1%	10.4%
Tern population in pairs	1500	1000	1200	1700
Gull population in pairs	700	800	1200	4000
Gull/tern ratio	0.5	0.8	1.0	2.4

Figure 25 gives the proportion of fish stolen by pirates in colony West 1967 in relation to the number of pirates present. It shows that an increase of the number of pirates coincides with an increase of piracy. However, we have no reliable figures of the number of pirates in the various years to check whether an increase of the gull population on the island resulted in an increase of the number of gulls involved in piracy. Besides, we should realise that a variety of circumstances which influence piracy, *e.g.* the age of the chicks, the size of the colonies watched, the black-headed gull's food situation, *etc.*, might have differed from year to year. Therefore, the conclusion that the growing gull population would have caused a relative increase of fish piracy should be regarded as tentative.

In 1966 clupeids (*Clupea sprattus* and *Clupea harengus*) and sand eels (*Ammodytes lancea* and *Ammodytes lanceolata*) made up 97% of the total number of fish brought into the colony. Other species were small in numbers and usually were refused by the chicks.

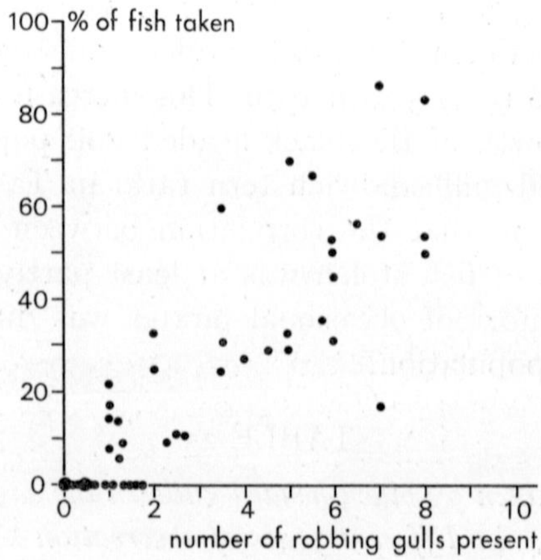

Fig. 25. Percentage of fish stolen by black-headed gulls in relation to the number of pirates (colony West, 1967).

Figures 26 and 27 give the length distributions for both groups of fish, together with the proportion lost as a result of piracy of black-headed gulls. Both figures show that for *Ammodytes* as well as for *Clupea* piracy highly increased with increase of length of fish. The same phenomenon appears if both groups of species are mutually compared: The relatively big sand-eels were stolen much more than the much smaller clupeids (32.5% against 15.5%).

It follows from these results, that fish-piracy, expressed as a percentage of numbers, fails to give us the right information on the amount of food taken by the gulls. In 1966, therefore, numbers have been converted to weight. (For details on method see I.2.4.2. and figure 14.) Table 26 gives the results. The difference in numbers and weight of the proportion of fish taken by the gulls reflects the preference of this pirate for relatively big fish.

If piracy of black-headed gulls influences chick viability, this can be expected to take place especially in a period of food shortage. It has been

Plate 9. A few days after the hatching of the first egg the chicks are lured away from the nest by the parents. The chicks in the photograph are between 5 and 10 days old; the second-hatched chick is notably smaller than the first. (phot. J. van de Kam).

10

11

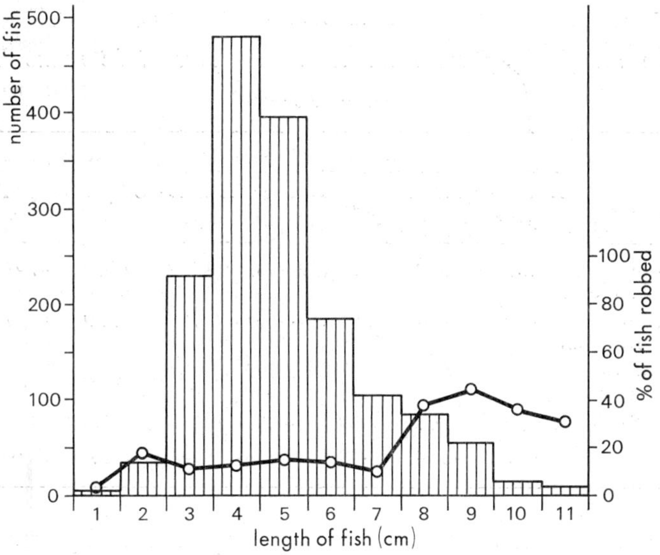

Fig. 26. Proportion of fish (*Clupea*) robbed by black-headed gulls from different length classes of the fish provisioned by the sandwich tern.

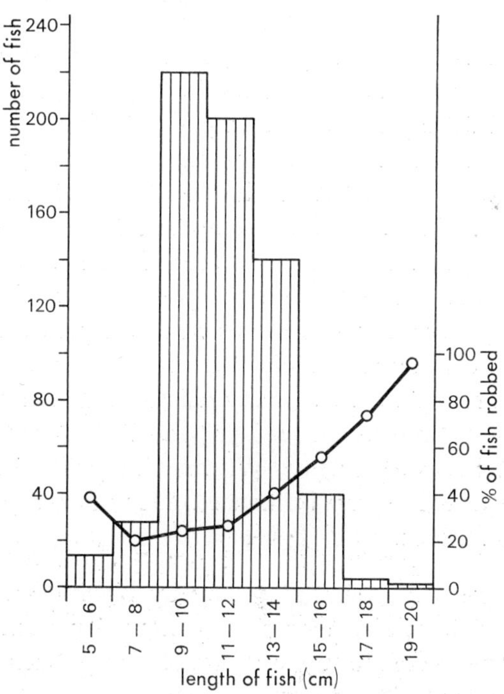

Fig. 27. Proportion of fish (*Ammodytes*) robbed by black-headed gulls from different length classes of the fish provisioned by the sandwich tern.

Plates 10 and 11. In reaction to Man, sandwich tern chicks may either crouch and keep motionless (plate 10) or huddle in large groups and walk away (plate 11).

TABLE 26

Piracy on the food of the sandwich tern expressed in numbers and in weight of fish

	Numbers			Weight in grams		
	Ammodytes	*Clupea*	Total	*Ammodytes*	*Clupea*	Total
fed to chicks	437	1336	1773 (79.6%)	1557	1144	2701 (65.8%)
robbed by black-headed gulls	160	202	362 (16.2%)	796	359	1155 (28.2%)
robbed by sandwich terns, common terns or arctic terns	50	44	94 (4.2%)	211	34	245 (6.0%)
total	647	1582	2229 (100.0%)	2564	1537	4101(100.0%)

shown that the chick's food supply is strongly reduced by strong winds (I.2.4.2.). This made it interesting to study black-headed gull piracy under different weather conditions.

Figure 28 gives the number of fish supplied at various wind-speeds in 1966, together with the proportion stolen by black-headed gulls. The proportion of sand-eels taken by the gulls strongly increased with wind-speed, but the proportion of clupeids taken by the gulls did not change with wind-speed in any important way. The differences between both groups of fish seemed related to the preference of the gulls for relatively big fish. In all weather conditions the sand-eels supplied were roughly of the same size. The size of the clupeids, however, decreased with increase of wind-speed. In stormy weather all clupeids were very small, which made them less attractive for the gulls (see also I.2.4.2.).

Under favourable weather conditions we never got the impression that chicks died as a consequence of food shortage. However, in 1966 in a 5-day period of stormy weather, many chicks died of starvation. Within this period, piracy removed between 50% and 100% of the daily weight of fish brought into the colony. Within the given period, therefore, piracy can be regarded to have influenced chick mortality in combination with weather conditions.

Summarizing we conclude that piracy by black-headed gulls on the food of the sandwich tern influences chick mortality when food gets short in a long lasting period of stormy weather. However, such periods are rare in the hatching season of the terns.

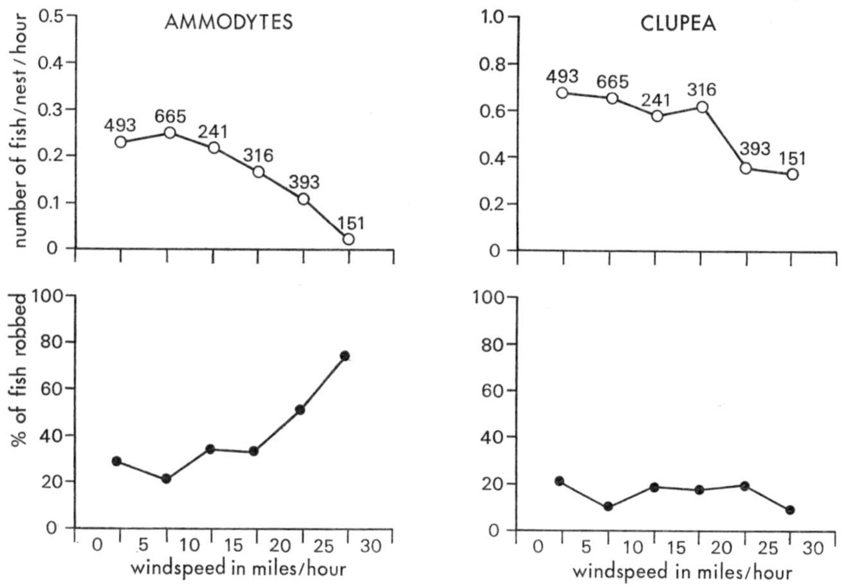

Fig. 28. Percentage of fish robbed by black-headed gulls in relation to wind-speed, compared with the number of fish supplied.

2.3. PREDATION BY HERRING GULLS AND GREAT BLACK-BACKED GULLS

2.3.1. Introduction.

Various authors (Brouwer, 1950; Cullen, 1960; Rooth, 1965) have mentioned herring gulls and great black-backed gulls as predators on the sandwich tern. Brouwer, citing the wardens of the Griend colonies, suggests that predation by great black-backed gulls on the brood of different species of terns might have been considerable in the seasons of 1925, 1926 and 1937, but detailed information is lacking.

Herring gulls and great black-backed gulls do not breed on Griend, but both species are common visitors to the island. At high tide they concentrate on the shell covered eastern banks. At low tide they are spread over the mud flats. Their numbers increase from a few hundred in May to some thousands in July and August.

2.3.2. Identification and occurrence of the predatory gulls.

By far the majority of large gulls that concentrate on the island at high tide always stayed outside the colonies of the sandwich terns and the other breeding birds. Only a few gulls took part in predation. Their numbers are given in Table 27. In all years of observation one adult herring gull seemed to have specialised in predation, whereas the other gulls seemed to rob occasionally. Although we were not completely sure that the predators which frequently visited the island within a certain year, were the same individuals every day, we have some evidence that most of them very likely were. An adult great black-backed gull, present for several weeks in 1970, had a conspicuous wing injury. This bird could easily be identified, just as the juvenile herring gulls and great black-backed gulls present on the island in 1970 and 1971. The adult herring gulls had no special markings, but they could be identified by their

TABLE 27

The number of predatory herring gulls and great black-backed gulls on Griend in six years of observation

	1965	1967	1968	1970	1971	1972
Herring gulls	1	2 (1)	1	3 (2)	1	2
Great black-backed gull				4 (3)	3 (3)	1

Numbers between brackets refer to gulls which incidentally hunted on the island.

individual hunting habits. There was no doubt, that, except for 1972, only one adult herring gull took part in predation each year and it seems even likely that the same individual returned to the island year after year.

The number of predatory herring gulls and great black-backed gulls present on the island was strongly influenced by inter-predator agression. In all years the central part of the island was the hunting territory of the adult herring gull. Other large gulls attempting to hunt in this area were immediately attacked and driven off by this bird.

Predation normally started in the second half of June, when the majority of sandwich tern eggs had hatched. It continued until the end of July when almost all chicks had fledged. In this period the predatory gulls could be watched at nearly any time of the day, but most of them were especially active at high tide and during the evening hours.

In the hunting behaviour of the predatory gulls two different types could be distinguished:

(1) From the air: In all years of observation one adult herring gull hunted on the island by flying over the colonies at a height of 5-15 m. As long as this predator had no success many flights were made in succes-

sion. Special attention was paid by this gull to the most densely populated parts of the island which usually were inhabited by sandwich terns and black-headed gulls. Now and then the predator intensified searching by lowering its height, sometimes swooping down to a distance of 2 m from the ground. This behaviour usually caused a large number of breeding birds, mainly black-headed gulls, to fly up from their nests. An egg or chick was caught by suddenly swooping down to the ground, picking up the prey in a split second and thereupon flying away to the beach.

(2) On foot: Several gulls approached the colonies on foot, walking in a straight line to the border. They stopped when the first breeding birds (mainly common terns and arctic terns) started attacking, without giving any further reaction than an occasional upward peck when the defender came too close. When, after a few minutes, these attacks had decreased the gulls resumed their walk towards the colony. This could be repeated several times until, finally, the gull had approached the colony up to a distance of 10 to 20 m. In this situation the predator might wait for a long time, until the majority of breeding birds had changed to normal colony life and the first chicks appeared. Thereupon the predator suddenly flew or ran to the nearest chick, eating the victim on the spot or taking it away to the beach of the island. Hunting on foot was usually confined to the end of the breeding season. It never took place in dense parts of the colony.

2.3.3. Predation.

Methods.

Data on predation by herring gulls and great black-backed gulls on the brood of different species of breeding birds have been collected by direct observation and pellet analysis.

In order to assess the total number of chicks robbed by the gulls, regular observations have been made in 1970, 1971 and 1972, Every day a general note was made with respect to the activity of each predator. On different days throughout the season, the predators were followed to assess their daily intake. Data on the frequency of presence of the predators, their daily intake and the ratio between the different species of chicks robbed, enabled an assessment of the total number of chicks of each species taken within a certain year. The calculation for 1970 will be given as an example: Between 8/6 and 28/7 1970 an adult herring gull regularly visited the colonies, whereas 6 other gulls appeared less regularly. In the beginning, the adult herring gull robbed a daily average of 7 chicks on 23 days (6 days of observation). On the average 4 chicks were taken on another 14 days (7 days of observation), which makes 217 chicks for this single predator. In a similar way it has been calculated that a total number of 68 chicks was robbed by the other gulls. So, altogether 285 chicks were robbed. For the sandwich tern this means that 61.2% (occurrence of sandwich tern chicks in the observations) of 285 = 174 chicks were taken by the gulls.

Results.

Predation on eggs.

Predation on eggs was seen in two years of observation: On 7/7 and 8/7 1965 an adult herring gull took 10-15 eggs or small chicks. Another 3 eggs were taken by an adult herring gull on 31/5 and 3/6 1972.

Since any predation attempt of large gulls always caused great disturbance among the breeding birds present, it is unlikely that many cases of egg-stealing have been overlooked. The above results, therefore, lead us to conclude that predation on eggs by herring gulls and great black-backed gulls cannot have influenced the breeding success of the sandwich tern in any important way.

Predation on chicks.

Predation on chicks of sandwich terns mainly occurred when the terns had spread over the island after having left the colony. Table 28 gives the estimated age of chicks taken by the large gulls in 1970, 1971 and 1972. It shows that the chicks were taken at any age. Usually, great black-backed gulls took bigger ones than did herring gulls. Small and middle-sized chicks, however, were preferred by both predator species.

TABLE 28

Age of sandwich tern chicks robbed by herring gulls and great black-backed gulls

Age of chicks in days	Observed number of chicks robbed 1970	1971	1972	Total
0- 7	6	–	6	12
8-14	5	5	6	16
15-21	5	2	2	9
22-28	5	1	2	8
29-35	3	–	1	4
36-42	–	–	1	1

All observations on chicks of different species of breeding birds robbed by herring gulls or great black-backed gulls are given in Table 29. Chicks of common- and arctic terns are taken together, because I could not distinguish between them. The category "unknown" refers to cases in which it was not possible to determine what prey species was taken. It follows, that the occurrence of chicks of sandwich terns in the diet of the gulls varied from year to year.

TABLE 29

Observed number of chicks of different breeding birds, robbed by herring gulls and great black-backed gulls

	1967	1968	1970	1971	1972
Sandwich tern			26	8	18
Black-headed gull	4		14	26	28
Common- or arctic tern	2	2	1		
Unknown	6	1	26	10	4
Total	12	3	67	44	50

With respect to the high number of chicks that could not be identified (category unknown), it seemed useful to verify the results of direct observation by another method. In 1970 and 1972 we collected a number of pellets dropped by the gulls and a comparison between the results of direct observation and pellet analysis is made in table 30. In both years, there was no significant difference in the ratio between the different species of chicks robbed, found by both methods (X^2 test on the data of the sandwich tern: 1970 — $p > 0.95$, 1972 — $p > 0.95$).

TABLE 30

Chicks of different species of breeding birds robbed by herring gulls and great black-backed gulls. (A comparison of the results, obtained by direct observation and pellet analysis)

	1970		1972	
	Observations	Pellets	Observations	Pellets
Sandwich tern	26 (63.4%)	18 (58.1%)	18 (36.0%)	5 (41.7%)
Black-headed gull	14 (34.2%)	11 (35.5%)	28 (56.0%)	7 (58.3%)
Common- or arctic tern	1 (2.4%)	2 (6.4%)	4 (8.0%)	0 (0.0%)
Total	41	31	50	12

Table 31 gives the total number of chicks of different species of breeding birds, robbed by herring gulls or great black-backed gulls in 1970, 1971 and 1972 (for calculation see methods). It follows, that the total number as well as the proportion of chicks of sandwich terns robbed by the gulls varied from year to year. (In Table 29 the data presented are not representative for the entire breeding season of a certain year. Therefore, the ratio between the different species of chicks given in Table 31 does not necessarily correspond with the ratio given in Table 29.)

TABLE 31

Total number of chicks robbed by herring gulls and great black-backed gulls

	1970	1971	1972
Sandwich tern	174	46	135
Black-headed gull	99	78	146
Common- or arctic tern	12	0	0
Total	285	124	281

With respect to the influence of predation on the breeding success of the terns it should be noted that herring gulls and great black-backed gulls take potentially viable chicks from the population. After the chicks have left the colony, mortality agents other than predation are of minor importance and the number robbed by the gulls will directly influence the numbers that fledge.

2.4. PREDATION BY OYSTERCATCHERS

Oystercatchers are known to prey upon the brood of colony breeding Larids. RUTTEN (1931) saw oystercatchers feeding many small tern chicks (*Sterna* spec.) to their own young and he mentions that only a few individual birds seemed to have adopted he predatory habit. KRUUK (1964) reported that one resident pair of oystercatchers ate 22 hen eggs laid out in a black-headed gull colony for an experiment. Predation on the eggs of the gulls, however, was not recorded. Comparable results were obtained by BROEKHUIZEN & DE SMIDT (1964) working in the sandwich tern colony on Griend: Oystercatchers took eggs of common tern, sandwich tern and black-headed gull laid out during experiments, but predation on the brood of these species was not observed in the natural situation.

Between 1965 and 1972 the breeding population of oystercatchers on Griend varied between 150 and 200 pairs. Nests were spread over the island and each year several pairs were recorded to incubate their eggs in the middle of the sandwich tern colonies. Throughout the season non-breeding individuals, in number varying between 500 and 2000, used the island as a high-tide roost. These birds never visited the sandwich tern colonies.

In the Griend colonies, predation on eggs (Plate 12) was observed in all years of observation. In general a case of predation seemed to be the result of an occasional encounter between predator and prey. Usually the oystercatchers hesitated for a long time before pecking at an unprotected egg and the majority of eggs that fell victim to this predator were eggs of 2-egg clutches already left by the adults and the first-hatched chick. On a few occasions specialists seemed to be active: In 1966 we got the impression that two different birds had specialised in stealing eggs during our visits to the main colony. If we approached this colony, both predators came running along from the beach, immediately pecking into an egg as soon as the terns flew up. When I did my work in the colony, they could damage several eggs in succession eating only a small part of the contents. On 6, 7, 12 and 14 June a total of 25 eggs was taken by these predators. To avoid this unpleasant influence of human disturbance, frequency and duration of our visits to the colony was minimized in the following years as described in II.1.2.2.

In 1970, 1971 and 1972 predation by oystercatchers on the eggs of the sandwich terns hardly occurred. Only one deserted egg was seen to be taken in 1970 and two in 1971. Predation on the tern chicks was never recorded.

2.5. PREDATION BY OTHER PREDATORS

In all years of observation several species of birds, which upon their diet can be regarded as potential predators of the sandwich tern, were recorded on the island. Their numbers are summarized in Table 32. All species mentioned can be regarded as predators on the brood, except for peregrine (predator on the adults) and arctic skua (pirate on the food). It follows from the table, that short-eared owl, buzzard, marsh harrier, hobby and merlin were only recorded in April, May or August, *i.e.* in a period in which predation on the brood was very unlikely or even impossible. Moreover, buzzard, marsh harrier, hobby and merlin were only seen passing the island. Grey herons also passed the island without showing any intention to land in the neighbourhood of the colonies. Ospreys sometimes caught a fish near the island and thereupon took a rest on the beach. Sparrow hawk, jackdaw and crow were usually seen on the island itself. However, these species always avoided the breeding birds present and never showed any intention to prey on the brood of the terns.

Hunting behaviour or predation attempts were only seen in the following species:

(1) Hen Harrier: On 25/6-1971 a hen harrier hunted for several minutes on the edge of the island. As far as I could ascertain no prey was taken.

(2) Peregrine: Three times a peregrine was seen hunting on the island. Waders and different species of terns including the sandwich tern were chased by these predators. Once a common tern was taken.

TABLE 32

The occurrence of a number of avian predators of the sandwich tern on Griend in seven years of observation (1965-1968 and 1970-1972)

Predator species	Potential prey	Total number of predator recorded (1965-1972)					
		April	May	June	July	August	Total
Blue heron	brood	–	5	–	14	9	28
Buzzard	brood	4	–	–	–	1	5
Honey buzzard	(brood)	–	11	–	–	–	11
Sparrow hawk	brood	–	2	–	–	–	2
Marsh harrier	brood	–	–	–	–	1	1
Hen harrier	brood	–	–	1	–	1	2
Osprey	(brood)	–	4	1	–	–	5
Peregrine	adults	1	1	1	–	–	3
Merlin	(brood)	–	1	–	–	–	1
Hobby	brood	–	1	–	–	–	1
Kestrel	brood	1	–	2	5	14	22
Arctic skua	fish	–	–	1	48	66	115
Short-eared owl	brood	3	3	–	–	–	6
Carrion crow	brood	1	4	1	–	–	6
Jackdaw	brood	17	12	–	1	–	30
Days of observation		87	217	210	210	84	

(3) Kestrel: In July and August kestrels were sometimes hunting on the island, hovering in the way typical for this species. On a few occasions a starling was caught. The hunting of the kestrels was always restricted to a small area around the beacon and the house of the wardens. In this area sandwich tern chicks seldom occur and predation on the brood of this species or any other tern species was never recorded.

(4) Arctic Skua: In all years of observation arctic skuas robbed successfully the food of the sandwich terns. Up to five birds have been observed actively involved in piracy at the same time. Usually, however, single birds were recorded. Piracy took place only at sea and lasted from a few minutes up to half an hour. Piracy by arctic skuas often took place in a period of stormy weather, *i.e.* a period of food shortage for sandwich terns. However, the irregular presence of only a small number of pirates and the relatively small number of fish taken made it unlikely that piracy by arctic skuas has influenced the breeding success of the sandwich tern.

Summarizing we conclude that the predator species mentioned in this section cannot have influenced the breeding success of the sandwich tern in any important way.

2.6. PREDATION BY MAMMALIAN GROUND PREDATORS

The only representative of the group of mammalian ground predators on Griend is Man. The hunting methods of this predator can be expected to be well understood and, therefore, need no further description.

In the nineteenth and the beginning of the twentieth century a large number of eggs was collected on the island every year (BROUWER, 1950). In 1916, however, Griend was given the status of a nature reserve. Ever since the eggs of the terns have been protected. Recently only a few eggs have been collected for scientific purposes.

It has been argued before, that the presence of human beings in the neighbourhood of sandwich tern colonies may easily cause disturbance, which in turn may lead to an increase of predation by black-headed gulls and oystercatchers (II.1.2.2.; II.1.2.3.; II.1.4.). In different years of observation my own visits to the colonies sometimes influenced predation in this way. As far as possible, however, this influence has been excluded from the predation results of the predators concerned.

2.7. THE PREDATOR SITUATION ON GRIEND COMPARED WITH OTHER COLONIES

In the literature information on predation in sandwich tern colonies is extremely scarce. In Ravenglass (Northumberland, England) where a few hundred sandwich terns nest in the middle of a large black-headed gull colony, foxes killed 12-15% of the adult terns during dark nights in 1962 (KRUUK, 1964). In the same area herring gull, carrion crow, stoat and hedgehog have been reported as predators on the black-headed gull. In Germany (DIRCKSEN, 1932; KÖNIG, pers. comm.) herring gulls and black-headed gulls are the main predators of the sandwich tern. In the German colonies mammalian ground predators are rare. However, they may be very harmfull. A striking example can be given from the island Norderoog (KÖNIG, pers. comm.). On this island mammalian predators are not endemic, but in 1947 a few rats immigrated. In 1948 and 1949 (the number of rats had increased enormously) all eggs and chicks of the sandwich terns (roughly a thousand pairs) were robbed. The rats also killed several adults. In 1949, during a high tide, the predators all drowned

In order to determine to what extent the predator situation on Griend can be regarded as representative for sandwich tern colonies in general, data have been collected on the occurrence of predators and the extent of predation in sandwich tern colonies in the Netherlands. The results of a number of colonies are presented in Table 33.

TABLE 33

The occurrence of predator species in the breeding areas of the sandwich tern in the Netherlands

NAME OF COLONY / PREDATOR	fox	polecat	stoat	weasel	hedgehog	rat	black-headed gull	herring gull	black-backed gull	common gull	oystercatcher	carrion crow
Rottumeroog	–	–	–	–	–	–	△	●	●	○	○	–
Griend	–	–	–	–	–	–	●	●	●	○	●	–
Texel schorren	–	–	△	–	△	○	●	●	●	△	○	–
Makkumerwaard	–	○	●	●	○	●	●	●	●	○	–	–
De Beer	–	○	△	○	△	△	●	●	○	●	●	△
Scheelhoek	–	–	△	△	–	●	●	●			○	–
Schouwen	–	–	–	–	–	△	●	△	△	△	○	–
Kwade Hoek	–	–	□	□	□	□	●	●	●	○	○	△
Hompelvoet	–	–	–	–	–	–	●	△	△	△	●	–
Rottumerplaat	–	–	–	–	–	–	△	●	△	△	○	–
Schiermonnikoog	–	–	–	–	□	□	○	○			○	
Veluwemeer	–	–	–	–	–	●	○					●

Symbols used: (-) not recorded, (△) recorded within a distance of a few kilometers, (o) recorded in colony area, (□) predation recorded on black-headed gulls, common terns or arctic terns nesting in the neighbourhood of the sandwich terns, (●) predation recorded on sandwich terns.

In the Dutch colonies fox and polecat are highly uncommon and predation on the sandwich tern has never been recorded. In a few colonies stoat and weasel are common in the neighbourhood of the nesting terns. In the Makkumerwaard colony both predator species are said to have preyed on the sandwich tern on a few occasions. Data, however, are not available. In the Kwade Hoek colony stoat and weasel occasionaly preyed upon the brood of the black-headed gull. According to the warden of this colony these predators never visit the sandwich tern colony.

Hedgehogs have been recorded from a number of colonies, but, once more predation on the sandwich tern has never been established. Hedgehogs, however, should be regarded as an important potential predator of the sandwich tern. On Schiermonnikoog, where the species has been introduced by Man, several hedgehogs have been found preying upon common tern eggs. According to the wardens the predators sometimes were so stuffed with eggs that they could hardly walk.

Rats have been recorded from several colonies. Very likely this species is the most harmful one of all mammalian ground predators, which might be related to its nocturnal feeding habits. According to the wardens of the Makkumerwaard and the Scheelhoek colonies rats took 'a large number of sandwich tern eggs in certain years'. In the Makkumerwaard groups of sandwich tern eggs and chicks gathered by rats were found in the vegetation surrounding the colony. On Texel rats are a common appearance in the black-headed gull colonies where they prey upon eggs and chicks. Once a rat killed several adult black-headed gulls on the nest (BOOT, pers. comm.). However, in the sandwich tern colony not far from these black-headed gulls, rats have never been recorded. A comparable situation is known from the Kwade Hoek, where rats frequently visit the gullery but do not enter the nearby sandwich tern colony.

With respect to the occurrence of avian predators it follows from the table that black-headed gulls, herring gulls and oystercatchers are common visitors to nearly all the colonies studied, black-backed gulls and common gulls are less common, while carrion crows have only been recorded from the rather inland Veluwemeer colony and from De Beer.

On the nature and the pressure of predation by avian predators the following information is available: Black-headed gulls prey upon eggs, chicks and fish in all the colonies where they breed in the company of the sandwich terns. Data on black-headed gull predation collected by ROOTH (1965) and BEIJERSBERGEN (unpubl.) are given in Table 34. It follows that the proportion of prey items taken varies from year to year and from place to place. My own data as presented in Table 21, 22 and 25 are roughly of the same order as those obtained by ROOTH and BEIJERSBERGEN.

Herring gulls and great black-backed gulls mainly rob chicks, but from a few colonies important predation on eggs has been recorded. Usually, only a small number of specialized predators is involved in predation. However, in a period of stormy weather, the number of predatory herring gulls and great black-backed gulls may increase considerably. In such a situation all the eggs and chicks in the sandwich tern colony on Texel (115 breeding pairs) were robbed in 1970. The same happened on the island Rottumerplaat (100 pairs) and on the Kwade Hoek (104 pairs) in 1972.

Predation by common gulls (on eggs and chicks) and oystercatchers (on eggs) is only reported from a few colonies. In all cases the number of eggs or chicks taken appeared to be small.

TABLE 34

Predation by black-headed gulls on fish, eggs and chicks of the sandwich tern in colonies in the Netherlands after Rooth (1965) and Beijersbergen (unpubl.)

Colony	Year	Predation by Black-headed gulls on: Fish	Eggs	Chicks
Texel	1961	10 %		
De Beer	1962	3 %	10-15 %	20 %
De Beer	1963	25 %	4 %	15 %
Hompelvoet	1970	10-30 %		
Hompelvoet	1971	23 %		
Kwade Hoek	1970	0-40 %		

It is not likely that the carrion crow has been an important predator of the sandwich tern in the colonies mentioned, except for the Veluwemeer colony. In this case predation might be a consequence of the rather inland location of the colony. Moreover, predation coincided with serious human disturbance which might have been of influence as well.

Summarizing we conclude that mammalian ground predators may be very harmful to the sandwich tern. However, in island colonies mammalian predators usually do not occur at all and in other colonies they appear to be rare. Extensive predation is only known from fox (England) and rat (Germany, Netherlands). Those mammalian predators which have been recorded in the neighbourhood of sandwich tern colonies (stoat, weasel, hedgehog, rat) seldom prey upon the sandwich tern. In a number of cases, however, these predators were recorded to prey upon the nearby common terns and black-headed gulls. Avian predators are common in all sandwich tern colonies. Black-headed gull, herring gull and great black-backed gull are the only species which have been recorded to be very harmful to the terns.

With respect to predation the situation on Griend can be regarded as representative for the situation in most of the colonies in the Netherlands.

2.8. THE INFLUENCE OF PREDATION ON THE BREEDING SUCCESS OF THE SANDWICH TERN

In the Griend colonies black-headed gulls, herring gulls, black-backed gulls, oystercatchers and arctic skuas have been recorded to prey upon the brood and to steal the fish of the sandwich tern. It has been shown that fish piracy of black-headed gulls in combination with adverse weather conditions may occasionally lead to chick mortality. Unfortunately, the influence of piracy on chick mortality could not be quantified. Since, in comparison to the overall influence of predation on breeding success, the influence of fish piracy seemed to be small, it has been excluded from the calculation given below. The extent of predation by oystercatchers and arctic skuas was so small that it was not possible to find an influence of these predators on the breeding success of the terns. Therefore, our treatment of the influence of predation on the breeding success of the sandwich tern will only refer to predation on the brood by black-headed gulls, herring gulls and black-backed gulls. Predation by herring gulls and black-backed gulls will be taken together as 'herring gull predation'.

Table 35 summarizes all data on predation by black-headed gulls and herring gulls collected in 1970 and 1971. The number of eggs laid and the number of chicks hatched have been calculated from nest checks and regular counts made in the colonies. The number of chicks fledged is

TABLE 35

Breeding results of the sandwich tern population on Griend in 1970 and 1971

Year	1970	1971
Number of breeding pairs	1742	2079
Number of eggs laid	2717	3626
Number of eggs lost:		
to black-headed gulls (viable eggs)	74	195
to black-headed gulls (non-viable eggs)	274	265
to other mortality factors	509	758
total	857	1218
Number of eggs hatched	1860	2408
Number of chicks lost:		
to black-headed gulls (viable chicks)	35	35
to black-headed gulls (non-viable chicks)	141	153
to herring gulls (viable chicks)	162	43
to herring gulls (non-viable chicks)	12	3
to other mortality factors (found dead)	132	379
to other mortality factors (not found)	382	593
total	864	1206
Number of chicks fledged	996	1202

calculated from the results of chick recoveries obtained during regular search trips covering the whole island in the period between hatching and fledging. Chick mortality, headed under 'other mortality factors' in the lower part of table 35, refers to mortality by diseases, starvation, chilling and drowning. With respect to the number of viable and non-viable eggs robbed by black-headed gulls the reader is referred to section II.2.2.3. and II.2.2.4.. The number of non-viable chicks robbed by herring gulls is calculated from the ratio of the number of chicks lost by 'other mortality factors' and the number of chicks surviving, giving full account of the fact that mortality by 'other mortality factors' strongly decreases with increase of age.

The influence of predation by black-headed gulls and herring gulls on the breeding success of the sandwich tern is derived from table 35 by calculating the increase of the number of chicks that would have fledged in absence of the predator concerned. Example: In 1970 in absence of black-headed gull predation $1860 + 74 = 1934$ eggs would have hatched. From the eggs hatched $1934/1860 \times (864 - 35) = 857$ chicks would have died. $1934 - 857 = 1067$ chicks would have fledged, which is 71 chicks more than the number of chicks that actually fledged.

For both years of observation the results are given in Table 36. It follows, that in 1970 the breeding success of the sandwich tern was influenced more by herring gulls than by black-headed gulls. In 1971, however, the influence of both predators was just the reverse.

TABLE 36

The influence of predation by black-headed gulls, herring gulls and great black-backed gulls on the breeding success of the sandwich terns in 1970 and 1971

	1970	1971
Eggs laid	2717 (100.0%)	3626 (100.0%)
Chicks fledged	996 (36.7%)	1202 (33.1%)
Calculated number of chicks fledging if predation by black-headed gulls is excluded	1067 (39.3%)	1351 (37.3%)
Calculated number of chicks fledging if predation by herring gulls and great black-backed gulls is excluded	1158 (42.6%)	1245 (34.3%)

Considering both years together, we conclude that the influence of predation by black-headed gulls and herring gulls on the breeding success of the sandwich tern was roughly the same. We also conclude that in both years predation was not a very important mortality factor of the brood of the sandwich tern in the Griend colonies.

3. ANTI-PREDATOR BEHAVIOUR

3.1. BEHAVIOUR IN RESPONSE TO BLACK-HEADED GULLS

3.1.1. Reactions of the sandwich terns.

Sandwich terns primarily defend their brood against black-headed gulls by sitting tight on eggs and chicks, thus avoiding exposure to predation. I never saw black-headed gulls chasing a sandwich tern from its brood, nor did I see a sandwich tern leaving eggs or chicks to attack a black-headed gull. When black-headed gulls approach the terns at close distance, either by foot or from the air, the terns threaten with open bill, crest erect and neck feathers ruffled, uttering a repeated 'gak gak gak'. In high intensity threat head and neck are moved up and down.

In 1970 data have been collected with respect to the distance at which incubating sandwich terns threatened black-headed gulls approaching the nest. Special paint markings were used to enable distances to be estimated. The results are given in Figure 29. It follows that the probability of an intruding black-headed gull eliciting a response strongly

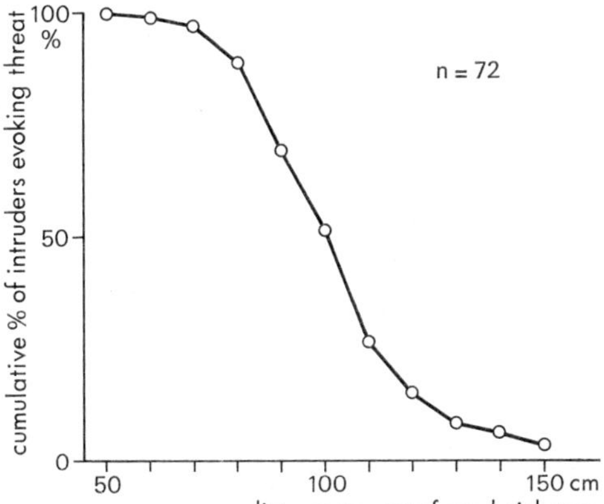

Fig. 29. The probability that a black-headed gull intruding on foot, elicits threat in a nest-owning sandwich tern, related to the distance between the gull and the nest.

increases between 1.5 and 0.5 m from the nest. The radius of the area defended averages 1 m. The figure does not account for the fact that the duration and intensity of threat strongly increases with decrease of distance.

3.1.2. Effects of the behaviour of the sandwich tern.

The behaviour of black-headed gulls responding to threatening sandwich terns differed from case to case. Usually the gulls reacted with an immediate change of direction, avoiding further approach. By far the majority of the gulls clearly avoided even coming within threat distance. However, experienced predators often roamed the terneries far within threat-distance and sometimes eggs were taken from exposed nests at less than half a meter of the incubating neighbours, apparently with complete ignorance of threat.

These observations suggest that against experienced predators threat is only effective at short distance. Consequently, a clutch temporarily left can be expected to gain protection only from neighbours present within short distance. To test this, in 1967 an experiment has been done.

The experiment consisted of 20 tests each of which was carried out in a different part of the colony. Per test 6 black-headed gull eggs were laid out: Four eggs were laid outside the colony at 2, 1½, 1 and ½ m, respectively, from the nearest nests in the colony edge. One egg was placed between two nests in the colony edge at a distance of 30-50 cm from both. One egg was laid inside the colony, surrounded by three or more nests within a distance of 30-50 cm. In each test the eggs were given a more or less similar position with respect to vegetation or other landscape features and the presence of breeding birds other than sandwich terns. Thus, egg-position was only different with respect to the presence of incubating sandwich terns.

Observations were made from a hide at a distance which did not influence the behaviour of the breeding birds present. Each test started after laying out the eggs and ended, either when all the eggs had been eaten, or after 3 hours of observation.

Figure 30 gives the order in which eggs were taken in relation to egg-position (first egg preyed upon was scored 1 and so forth). If one egg survived the experiment, it was consequently scored as 6. If two eggs were left, each was given the same score (of 5.5) and so on. In a few occasions oystercatchers robbed the eggs, which is indicated in the figure. The figure shows, that eggs outside the colony had equal chances to be robbed. However, the outside-colony eggs were preferred above eggs in the colony-edge and the latter above eggs inside the colony. The eggs outside the colony were normally eaten immediately after detection. On the eggs in the colony-edge several unsuccessful predation attempts were made before predation actually took place. Eggs inside the colony

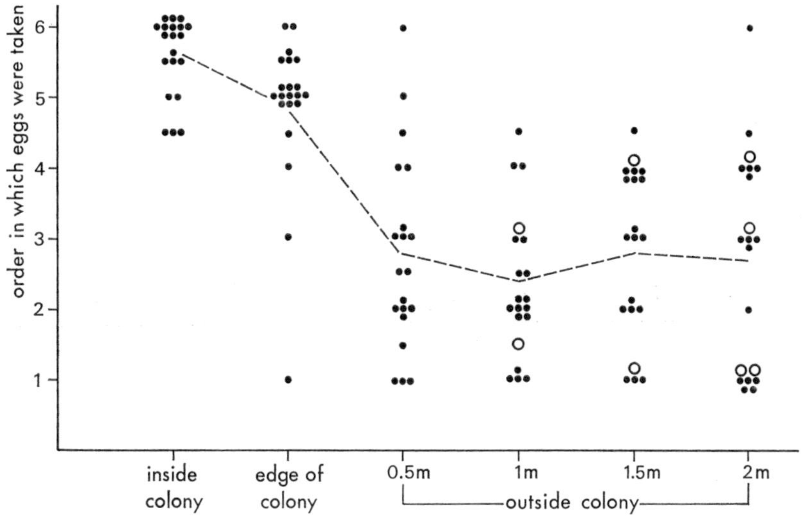

Fig. 30. Order in which eggs, placed at different positions in the sandwich tern colony, were taken by black-headed gulls. A few eggs were taken by oystercatchers (open dots). The broken line connects median preferences. Further explanation in text.

often remained untouched till the end of the experiment, despite numerous predation attempts. If an egg disappeared from the middle of the colony, this always took place when several gulls made an attempt at the same time. Single gulls were successfully kept at a distance by the upward pointed bills of the terns.

These results clearly indicate that a clutch temporarily left gains protection from neighbouring nest-owners. For a particular nest, the amount of protection received seems determined by the number of neighbours by which it is surrounded as well as by the distance of these neighbours. The experiment suggests that neighbouring nests give only protection within a distance of 50 cm.

3.1.3. Reactions of common terns and black-headed gulls.

It has been shown that sandwich terns primarily defend their brood against black-headed gulls by sitting tight on eggs and chicks, avoiding exposure to predation. In contrast, the reactions of common terns and black-headed gulls to this predator are characterised by flying up, attacking and chasing (Plate 17).

Figure 31 gives the relation between the proportion of predation attempts by black-headed gulls evoking attack in nest-owning common

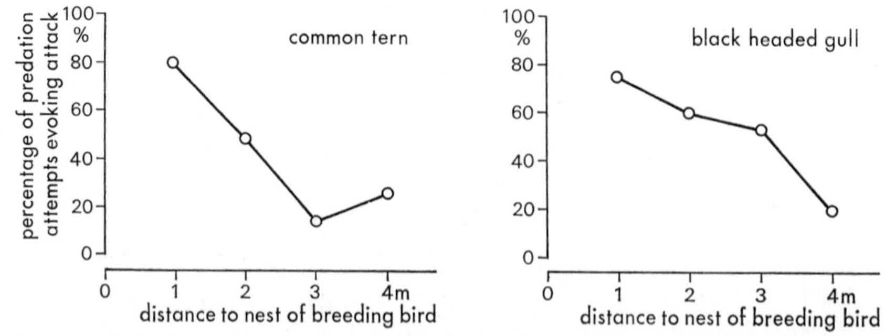

Fig. 31. Percentage predation attempts of black-headed gulls on eggs placed at different distances from the nests of common terns and black-headed gulls evoking attack in the nest-owning birds.

terns and black-headed gulls. The data have been derived from a series of tests (21-30 May 1966) in which eggs were laid out at different distances from common terns and black-headed gulls nesting in the neighbourhood of sandwich tern colonies (for details on methods see II.2.2.2. and Table 19). The figure shows, that the robbing gulls were frequently attacked by the nesting common terns and black-headed gulls, but in both cases the number of attacks strongly decreased between 1 and 4 m from the nest.

In the above tests the attacks of common terns and black-headed gulls were always immediately followed by flight in the predators. These attacks, therefore, proved to be highly effective in driving off predatory black-headed gulls. Sandwich terns nesting within 'attack distance' of either common terns or black-headed gulls can thus be expected to gain protection from these neighbours.

3.1.4. The influence of the anti-predator behaviour of common tern and black-headed gull on the hatching success of the sandwich tern.

To investigate the possible effect on the hatching success of the sandwich tern of the anti-predator behaviour of common terns and black-headed gulls nesting in their neighbourhood, I selected in 1967 three sandwich tern colonies in which either common terns or black-headed gulls were present. Within these colonies sandwich tern nests were checked for success as described in II.1.2. In Table 37 these nests have been grouped with respect to their distance to the nearest nest of a common tern or black-headed gull; for each group the predation data are given. It follows, that the proportion of sandwich tern eggs preyed

TABLE 37

Predation on eggs of sandwich terns nesting at different distances from common terns and black-headed gulls

Distance to nearest nest of Common Tern	Number of nests of Sandwich Tern	
	Total	Preyed upon
0-1 meter	36	4 (11.1%)
1-2 meter	60	12 (20.0%)
2-3 meter	87	29 (33.3%)
3-4 meter	31	13 (41.9%)
>4 meter	28	14 (50.0%)
Total	242	72 (29.8%)

Distance to nearest nest of Black-headed Gull	Number of nests of Sandwich Tern	
	Total	Preyed upon
0-1 meter	10	3 (30.0%)
1-2 meter	26	14 (53.8%)
2-3 meter	13	9 (69.2%)
Total	49	26 (53.1%)

upon was negatively correlated with the distance to the nearest nest of a common tern or black-headed gull.

Before accepting these results it is necessary to take account of the fact that clutch size, synchronization of egg-laying and nest position can be of influence on the hatching success of sandwich terns (Chapter II.1.). As the groups of nests studied were not different with respect to clutch size and synchronization, these factors cannot have influenced the above results. However, differences in nest position were likely to occur between the different groups of nests. Consequently this factor may have been of influence. Unfortunately it was not possible to quantify this influence, and this made me decide to do the following experiment:

On 29 and 30 May 1967 rows of 3 single black-headed gull eggs were laid out in places where sandwich terns nested in the neighbourhood of common terns or black-headed gulls. Within each row the eggs were placed at different distances from the nest of either a common tern or a black-headed gull (1, 2 and 5 meter). The distance of the eggs to the nearest sandwich tern nest was always one meter. With respect to vegetation and other environmental factors all the eggs had a similar position.

One and two hours after the eggs had been laid out, a check was made to find which eggs had been robbed by black-headed gulls. The results are given in Table 38. It follows from the table that the proportion of eggs robbed was positively correlated with the distance at which they had been placed from a black-headed gull's nest. Of the eggs placed near the nests of common terns, the proportion preyed upon at a distance of 1 and 2 m was not different, but that of eggs preyed upon at 5 m was much higher (for statistical analysis the number of eggs laid at 1 and 2 m have been taken together and compared with the eggs laid at 5 m: common tern X^2 test, $p < 0.05$; black-headed gull X^2 test, $p < 0.025$).

TABLE 38

Predation on eggs laid out alongside the sandwich tern colony at different distances from the nests of common terns and black-headed gulls

Distance egg-nest in meter	Species of incubating bird	Number of eggs presented	Number of eggs robbed after 1 hour	2 hours
1	Common tern	12	5	7
2	Common tern	12	5	7
5	Common tern	12	9	11
1	Black-headed gull	12	3	3
2	Black-headed gull	12	5	5
5	Black-headed gull	12	8	9

The observations on the attack-frequency of common terns and black-headed gulls on black-headed gulls preying at different distances from their nests (Fig. 31) strongly suggest that common terns and black-headed gulls only protect sandwich terns which have their nest within a distance of 4 m. To investigate which proportion of the sandwich tern population in different years of observation nested within this protected 4 m distance, inter-nest distances have been measured in 1966, 1967, 1970 and 1971. Table 39 gives the results. It follows that in 1966 and 1967 the majority of sandwich terns nested near common terns, whereas in 1970 and 1971 the majority of terns nested near black-headed gulls. An increase of the breeding population of black-headed gulls might have caused this phenomenon. However, between the different years the total number of sandwich terns having their nest within a distance of 4 m from either a common tern or a black-headed gull differed only slightly. Roughly 90% of the sandwich terns nested within the protective four meter distance of one of both species in all years of observation.

Summarizing we conclude that the brood of the sandwich tern is safe from black-headed gull predation as long as the adults stay on the nest. A clutch temporarily left gains protection from neighbouring sandwich terns nesting within a distance of 50 cm. Nesting common terns and black-headed gulls give protection against predatory black-headed gulls to sandwich terns present within a distance of 4 m. This protection increases with decrease of distance.

TABLE 39

Nest position of sandwich terns relative to common terns and black-headed gulls

Year	Number of nests checked	Number of nests of Sandwich Tern within 4 m of Common Tern	Black-headed Gull	One of both
1966	918	599 (65.3%)	551 (60.0%)	881 (96.0%)
1967	405	283 (69.8%)	206 (50.9%)	380 (93.8%)
1970	182	61 (33.5%)	151 (83.0%)	157 (86.3%)
1971	576	291 (50.5%)	446 (77.4%)	521 (90.5%)

3.2. BEHAVIOUR IN RESPONSE TO HERRING GULLS AND GREAT BLACK-BACKED GULLS

3.2.1. Reactions of the sandwich terns.

On the nest.

The reactions of the sandwich tern to an appearing predator are manifold and depend on the behaviour of the predatory gull and some properties of the colony, such as the number of birds and their breeding phase.

When a predator appeared, flying fast and at a height of more than 5 m, the terns did not show a reaction at all. When the gulls came nearer, they might raise their head and neck, sleeken their plumage and utter alarm calls, which sounded like a yawning 'wah', and which were performed with the neck stretched and the bill pointed obliquely upward. Sometimes a predator stalled above the colony, or swooped down, both clear indications of hunting. In this situation many birds would stand up from their nests, raise their wings and either the whole colony or a part of it might fly up. Usually the predatory gull continued on its way without mingling with the flying birds. Individual terns never separated from the flock to attack or to pursue the gull and it was always striking that they alighted so soon.

When a gull made a real attempt to alight, it often disappeared among the flying birds and was difficult to follow. A small number of terns might make faint diving attacks, but furious attacks or long-lasting pursuits never occurred.

In a few occasions we observed that a herring gull landed in the colony. This always made all the neighbouring terns fly up and concentrate in a dense flock above the predator, making an impressive amount of noise. Regular diving attacks of varying intensity were made, but the number of birds taking part in it was always small. They rarely approached the gull within a distance of 1 m. No pecking or hitting of the predator ever took place.

I have never seen predatory herring gulls or great black-backed gulls standing in the colony for more than a few seconds, but in 1965 there were some sick herring gulls, which stranded in the sandwich tern colony, where they stayed for a rather long time. In these cases the attack frequency strongly increased during the first 30 seconds, after which it reached a more or less constant level. A few minutes later this frequency dropped again and many birds which had their nest at a distance of more than 10 m, alighted.

The tendency of the terns to fly up from the nests was much lower in large colonies than in small ones. On 30/5 and 3/6 1972 a comparison was made between the reactions of breeding birds in different colonies. The results are presented in Table 40. Colony South consisted of several groups of nests of more than a hundred each. This colony is taken together with colony West G (one group) for statistical analysis and compared with colony West K which consisted of some small and scattered groups

TABLE 40

Reactions of nesting sandwich terns to a hunting herring gull flying over the colony

Name of colony	Total number of observations	Observations in which the terns reacted with flying up	Observations in which the terns stayed on their nests
South (1200 nests)	16	0	16
West G (129 nests)	16	0 (1)	15
West K (26 nests)	21	8 (6)	7

The figures between brackets refer to observations in which part of the colony flew up. For statistical analysis they are taken together with the 'flying up' group.

of nests. The difference is highly significant (X^2 test, $p < 0.005$).

On the same days three nests were robbed in colony West K, whereas nothing was taken from the other two colonies. Inside and near colony West G several chicks of black-headed gulls were taken. Table 41 gives the reactions of the terns in this colony in a number of cases when predation took place. It follows that they only flew up when the gull landed in the middle of the colony or at a few meters from the edge.

TABLE 41

Reactions of sandwich terns to a herring gull alighting at different distances from the nests

Landing position of herring gull in relation to nearest nests in colony-edge	Number of breeding birds flying up
8 meter	none
5 meter	none
4-5 meter	none
3 meter	10
colony-edge	50
colony centre	many
colony centre	many
colony centre	all

After nest departure.

After leaving the nest the sandwich tern chick turns to a more cryptic life and hides itself in the vegetation. The adults show a striking increase in aggressiveness. In contrast with the situation in the colony, they are much more inclined to fly up and to attack a predator.

If a predatory herring gull or great black-backed gull flew over an area where large chicks were present, this always caused the terns to fly up when the predator was still at a distance of more than 50 m. Individual terns pursued the predator over a distance varying from 10 to 100 m. Attacks began with a horizontal approach at high speed from behind. Attacks never resulted in bodily contact. At a distance of 50 cm to several meters from the predator the attacker suddenly veered off and up.

If the predator was present on the ground, the terns concentrated in a sparse flock above the marauder. Depending on the number of terns present several attacks might be made. Attacks consisted of aerial swoops which often came very close. Above the predator the terns usually made pecking movements. However, actual contact between attacker and predator was not recorded with certainty.

3.2.2. Effects of the behaviour of the sandwich tern upon the predators.

The above description of the reactions of the sandwich tern shows that the species has two different mechanisms to defend the brood against herring gulls and great black-backed gulls. On the nest the terns tend to sit down, hiding eggs and chicks from the view of the predator. After nestleaving the terns tend to fly up, to attack and to chase the ennemy.

Observations have shown that eggs and small chicks were only taken from small colonies or scattered groups of nests, whereas nothing was taken from large colonies. These differences occurred despite the fact that the gulls did hunt in large colonies (*e.g.* in 1972 several black-headed gull chicks were taken from the middle of a large sandwich tern colony, which clearly indicates that the gulls do hunt and are able to land in a large sandwich tern colony). Since the tendency to stay on the nest is relatively low in the smaller colonies, differences in predation success between large and small colonies can only be explained by differences in prey vulnerability which, in turn, is dependent on whether or not the nest-owner leaves the nest. Hiding the brood from the view of predators by sitting on it, thus seems a very effective defense.

The effect of attacks and chases in the air was difficult to establish. On Griend the presence of herring gulls and great black-backed gulls always elicited reactions of different species of breeding birds and the effect of attacking and chasing of sandwich terns, therefore, could not be separated from the combined effect of the reactions of the different species. We shall go further into this matter in the following sections.

3.2.3. Experiments with stuffed and living herring gulls.

3.2.3.1. *Introduction.*

The behaviour of common terns and black-headed gulls responding to predatory herring gulls and great black-backed gulls markedly differed from the behaviour of the sandwich terns. Throughout the whole breeding season common terns and black-headed gulls were highly inclined to fly up from the nest when a predator appeared. Moreover they attacked a gull furiously and chased it over a long distance.

In the Griend situation, sandwich tern, common tern and black-headed gull always nested together and a predation attempt by herring gulls and great black-backed gulls usually evoked reactions from different species of breeding birds almost simultaneously. In general the predatory gulls were wrapped in a whirling flock of birds in which it was impossible to study to what extent each species contributed to the defense of the colony area. Moreover, the reactions of the different breeding birds strongly depended on the behaviour of the predator, which in turn was difficult to follow. For this reason, I decided to test the reactions of the breeding birds in a more standardized situation by presenting stuffed and living herring gulls in the colony. The aim of these experiments was (1) to study qualitative and quantitative differences in anti-predator behaviour of sandwich terns, common terns and black-headed gulls and (2) to obtain evidence whether or not sandwich terns gain protection from neighbouring common terns and black-headed gulls against predatory herring gulls and great black-backed gulls.

3.2.3.2. *Material and methods.*

Stuffed herring gulls in adult plumage were used in the experiments. Handreared herring gulls in juvenile as well as in adult plumage were used to compare the reactions of the different breeding birds to a stuffed and a living predator model.

KRUUK (1964), working with stuffed predator models in a black-headed gull colony, has listed the various practical difficulties which are involved in the presentation of predator models in a colony. KRUUK tested different methods of mechanical presentation, but he finally preferred to set the model out himself, thereupon watching the reactions of the breeding birds from some distance outside the colony. I mainly followed the same method.

On Griend the experiments were all done on the beach-side of the island, as this was the only place where the colonies had a sharply defined edge. After the models had been put in position (the living gulls were tethered to a peg) I walked to a previously prepared observation post where a pair of binoculars had been placed on a tripod. This post was always situated more than 100 m from the edge of the colonies. Observations started one minute after I had left the colony. Control-observations showed that within this period the breeding birds had always resumed sitting.

Observations on the number of attacks made by the different breeding birds on the model (Plate 16) were recorded on tape. Every swift approach, followed by a withdrawal from the model, was counted as an attack, whether it resulted in actual contact or not. Simultaneously with my observations on the attacks, a second observer made a note on the number of birds concentrated above the model. The experiments were all done in 1967 and 1968.

3.2.3.3. *Analysis of the reactions of common terns, black-headed gulls and sandwich terns.*

In 1968 an experiment was undertaken to study the reactions of common terns, black-headed gulls and sandwich terns to a stuffed herring gull placed at different distances from the nests. As it was our aim to study the reactions of each species separately, more or less 'unmixed' parts of the colony were selected for the experiment. At each location chosen one series of 5 min. tests was made, respectively inside the colony and at different distances from the edge. The intervals between the tests of each series were one day.

Description of the reactions.

Depending on the distance of the model to the nearest nests, all sandwich terns nesting within a certain colony area either alighted on their nests, concentrated above the model or flew around in wide circles. Sometimes, flying around resembled the formation flights which are shown in reaction to the larger raptors (II.3.4.1.). If concentrated in the air a relatively small number of birds was involved in attacking the model. Attacks usually consisted of faint aerial swoops or fast horizontal flights at a few meters above the model. However, if the model was very close to the nests, the terns might approach it very closely. Above the model they often made pecking movements, but actual contact rarely occurred.

In contrast with the sandwich tern, common tern and black-headed gull reacted much more individually to the model. Depending on the distance between model and nest a varying number of nest owners concentrated in a flock above the predator. Usually the model was continually attacked by a relatively large number of individuals. The attacks of common terns consisted of aerial swoops from varying height which often came very close to the model. Above the model the terns often stretched out their feet, made pecking movements and sometimes defecated. Very often the model was pecked at in head and neck. The black-headed gulls usually attacked the model by aerial swoops which started up to a few meter above the model. Sometimes these swoops had a horizontal approach. If the model was close to the nests, the black-headed gulls might steeply swoop down from a hovering position at 1-2 m above the predator. The model was often hit with the feet or even with the entire body, and sometimes it was pecked at.

Attack-frequency and size of the area defended.

Fig. 32 gives the frequency of attacks of common terns, black-headed gulls and sandwich terns on the stuffed gull when placed at different distances from the nests. For each species the frequency of attacks strongly decreases with increase of the distance to the model. The spread in the data is due to a local difference in the number of birds present at the same distance from different parts of the colony. Following KRUUK

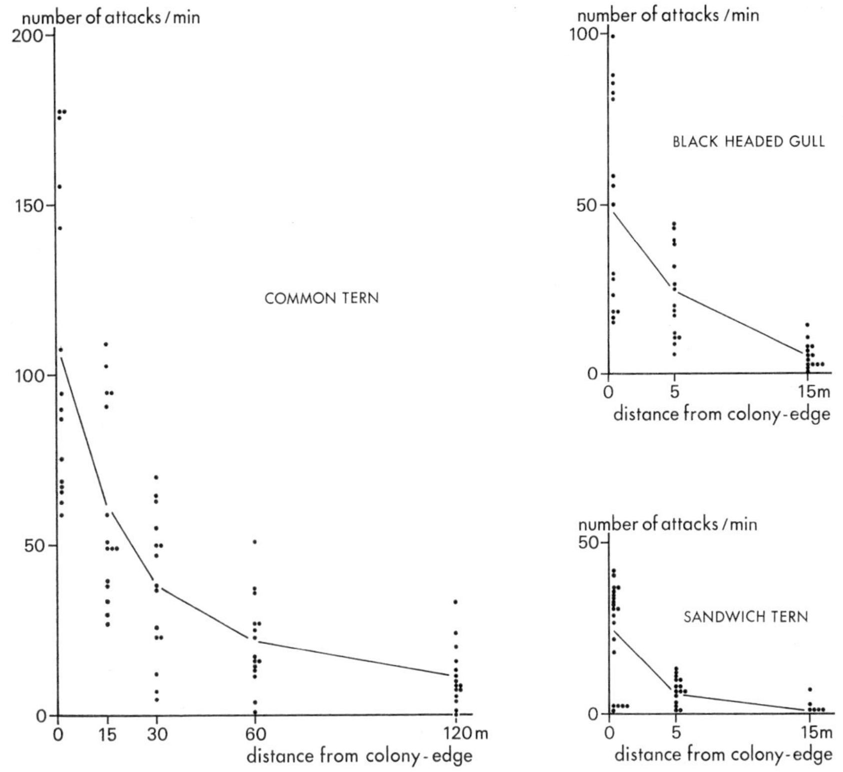

Fig. 32. Frequency of attacks of common terns, black-headed gulls and sandwich terns on a stuffed herring gull placed at different distances from the nests.

(1964), I have taken the distance at which the birds attacked the model half as frequently as they did near their nests as an indicator of the size of the area over which the predator was attacked. Then the ratio of the radii of the response area of common tern, blackheaded gull and sandwich tern is 8:2:1, the surface of these areas being 64:4:1.

Relative attack-frequency.

The relation between the frequency of attacks and the number of birds concentrating above the model is given in Fig. 33. It follows, that the relative attack-frequency strongly decreased in the order common tern — black-headed gull — sandwich tern. The overall averages respectively were 4.38, 1.24 and 0.26 attacks per bird per minute.

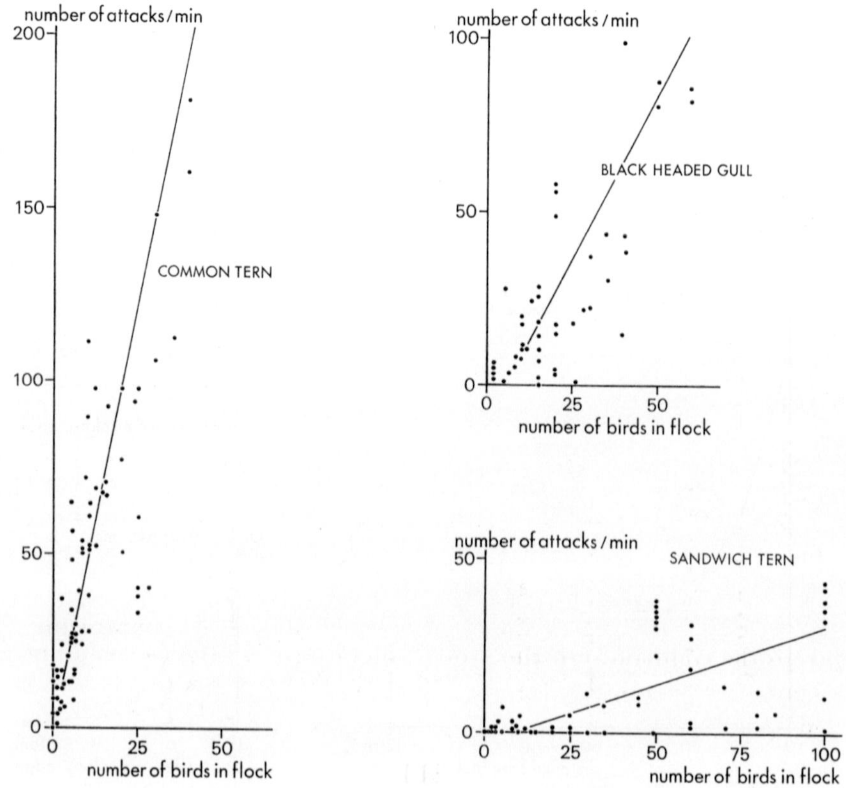

Fig. 33. The relation between flock size and attack frequency of common terns, black-headed gulls and sandwich terns responding to a stuffed herring gull placed in the colonies.

Proximity of attacks.

In the tests in which the stuffed herring gull was placed inside the colony, the proximity of the attacks on the model was estimated. A height of 50 cm above the back of the model was chosen to classify the swoops into 'nearby' and 'distant' attacks. The results are presented in Table 42. It shows, that the proportion of nearby attacks slightly decreased in the order common tern — black-headed gull — sandwich tern.

Changes of anti-predator behaviour throughout the breeding season.

In order to compare at different times of the season the reactions of common terns and sandwich terns to a stuffed herring gull, a number of five minute tests made on 22 May at 60 and 15 m, respectively, from the edge of the common tern colony (3 tests) and the sandwich tern colony

TABLE 42

Proximity of attacks of common terns, black-headed gulls and sandwich terns on a stuffed herring gull placed inside the colony of the different species

	Number of attacks Total	Within 50 cm from model
Common Tern	1611	1010 = 62.7%
Black-headed Gull	696	373 = 53.6%
Sandwich Tern	494	250 = 50.6%

(4 tests) have been repeated at exactly the same places on 17 June. Common terns and sandwich terns both incubated their eggs on 22/5, whereas they had chicks on 17/6. The results are presented in Table 43. On 17/6 for common terns as well as for the sandwich terns there was an increase of the number of birds concentrated above the model, of the absolute attack frequency and of the relative attack frequency. In the sandwich tern the proportion of nearby attacks also had increased considerably on 17/6. In the common tern the proportion of nearby attacks could not be measured because of the exceptionally high attack frequency.

TABLE 43

Reactions of common terns and sandwich terns to a stuffed herring gull presented at different times of the breeding season

	Common tern		Sandwich tern	
	May 22	June 17	May 22	June 17
Average number of attacks/min.	21.2	61.1	0.0	16.0
Average number of birds in flock	6.7	12.3	5.5	32.9
Average number of attacks/bird/min.	3.18	4.95	0.13	0.76
Proportion of attacks within 50 cm	58.5%	high	28.6%	62.5%

To minimize the disturbing influence of the experiments with the stuffed gull, the above tests have not been done with respect to the black-headed gull. KRUUK (1964) has shown that in this species the attack-frequency on man reaches a maximum in the period in which the eggs hatch. This is in agreement with the situation in common tern and sandwich tern.

Anti-predator behaviour in reaction to a living herring gull.

During the experiments the stuffed herring gull sometimes fell over as a result of the attacks of common terns and black-headed gulls. This always caused a considerable increase of the intensity and the frequency of the attacks. It looked as if the movement of the model was the direct cause of the change of behaviour of the breeding birds present.

To examine the possible influence of predator movements on the reactions of the breeding birds an experiment has been done by placing a living herring gull in the common tern colony. Six tests, each lasting 30 min., were made at different locations with intervals of half a day. The results are given in Figure 34. On average the attack-frequency during test 1 was much higher than during the tests 2 to 6. During test 1 the herring gull repeatedly walked around and sometimes fluttered its wings. During the other tests, however, the gull usually sat down and did not move in any important way. In all cases when the gull started to move several terns flew up from their nests joining the flock above the predator. At the same time the number of attacks made strongly increased, while the distance between the attacking birds and the gull decreased. (During test 1 the gull was 17 times pecked, whereas it was never touched during the other tests.) The relatively high attack-frequency during test 1 thus seemed a direct consequence of the movements of the predator.

To minimize disturbance the above tests have not been done in the colonies of black-headed gulls and sandwich terns. As black-headed gulls and sandwich terns showed an increase of aggressive reactions to a stuffed model falling over, it is very likely that these species react to movements of a predator in the same way as described for the common tern.

Comparison of the reactions of common terns and sandwich terns to a stuffed and a living herring gull.

To compare the reactions of common and sandwich terns to a stuffed and a living herring gull when simultaneously presented, these models have been placed at six different locations near the colonies of common tern and sandwich tern at 30 and 60 m from the edge respectively. Each test lasted 10 min. The results are given in Table 44. It follows that the stuffed model was attacked much more often than the living model. Usually the living gull immediately sat down on the ground and remained motionless for the rest of the test. Once (test 4 in the common tern

Plate 12. A black-headed gull and an oystercatcher quarreling over an egg stolen from the sandwich terns.
Plate 13. A black-headed gull eating from the remains of a sandwich tern egg. (phot. J. van de Kam).

12

13

colony) when the living gull was moving from time to time, it was attacked much more frequently than the stuffed gull.

This experiment once more shows that movements of the predator model are immediately followed by an increase of the number of common

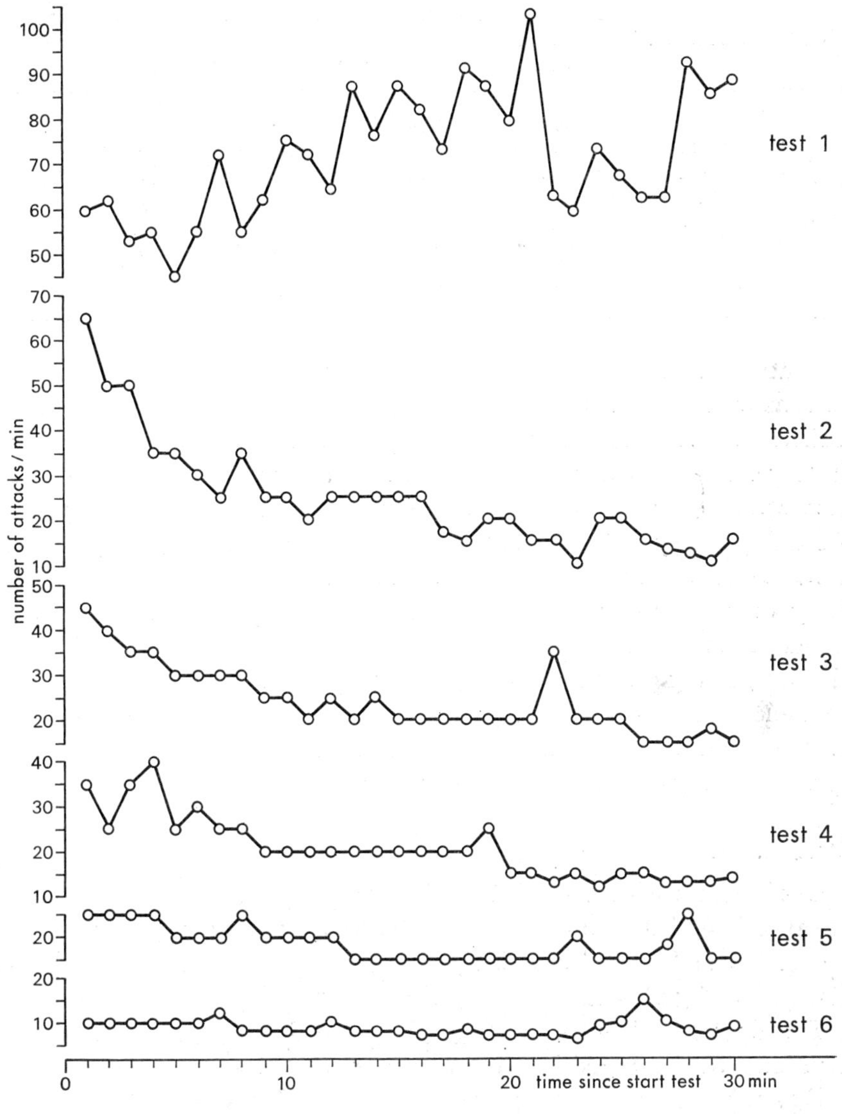

Fig. 34. Attack rate of common terns confronted with a living herring gull tethered in the colony.

Plate 14. A black-headed gull chasing a fish-carrying sandwich tern.

TABLE 44

Frequency of attacks of common terns and sandwich terns on a stuffed (adult) and a living (juvenile) herring gull presented simultaneously near the colony

Test no.	Attacks of Common terns on Herring gull		Attacks of Sandwich terns on Herring gull	
	Stuffed	Alive	Stuffed	Alive
I	490	246	9	9
II	105	30	70	16
III	52	22	118	32
IV	306	467	41	17
V	148	75	49	27
VI	135	134	20	6
Total	1236	974	307	107

tern attacks. However, when the living gull did not move, common terns as well as sandwich terns more often attacked the stuffed one. The stuffed gull and the living one mainly looked different in posture and plumage. The stuffed gull was fixed in standing position with head and neck stretched, the open bill pointing upwards. This rather aggressive posture (anxiety upright, TINBERGEN, 1959) which markedly differed from the hunched posture of the living gull sitting on the ground, very likely was the cause of the difference in attack frequency evoked by both models. It is not likely that differences in plumage (stuffed gull in adult plumage, living gull in juvenile plumage) played an important role, for both adult and juvenile herring gulls have been recorded to prey upon the terns and I never found any important difference in the reactions shown by the breeding birds to both plumage types.

Conclusions.

Experiments with a stuffed herring gull have shown that the size of the area defended around the nest, the relative attack frequency and the proportion of 'nearby' attacks, decrease in the order common tern — black-headed gull — sandwich tern.

The aggressive reactions of the breeding birds mentioned reach a maximum in the period in whcich the eggs hatch.

Experiments with a living herring gull have shown that movements of the predator, and very likely also its posture, influence the number of attacks made by common terns, black-headed gulls and sandwich terns.

The reactions of common terns and sandwich terns to a stuffed and a living herring gull were only different with respect to frequency and intensity. The reactions of the different species of breeding birds to a herring gull hunting under natural circumstances usually changed from one moment to the other, depending on the behaviour of the predator. However, the reactions of the breeding birds to a predator hunting under natural circumstances were not basically different from the reactions shown to the stuffed and the living gull used in the experiments. This convinced me, that it is permitted to extrapolate the results with predator models to the natural situation.

3.2.3.4. *Combined attacks of common terns, black-headed gulls and sandwich terns on a stuffed herring gull presented in the sandwich tern colony.*

It follows from the analysis of the anti-predator behaviour of common terns, black-headed gulls and sandwich terns that in sandwich tern colonies mixed with common terns and black-headed gulls both latter species can be expected to play a major role in defending the communal breeding area. To test this for the Griend situation an experiment has been done between 27 May and 6 June 1967. At eight different locations randomly chosen from the sandwich tern colony a stuffed herring gull was placed inside the colony and at 15 m outside the colony edge. Each location was used once for an inside-colony test and once for an outside-colony test. Tests inside and outside the colony were alternately done with intervals of two days. Each test lasted five minutes. The number of common terns and black-headed gulls nesting among the sandwich terns varied from one place to the other.

Table 45 gives the average number of attacks per minute on the model made by the different species of breeding birds. On the average the common terns showed more attacks than the black-headed gulls, the black-headed gulls more than the sandwich terns.

The number of attacks by black-headed gull and sandwich tern strongly decreased with increase of the distance between colony-edge and model. But in the common tern the number of attacks went up with increasing distance between colony-edge and predator-model, in contrast with what one might have expected. However, observations showed that the number of attacks made by the common terns was strongly and negatively influenced by the number of attacking black-headed gulls and sandwich terns present. An increase of the total number of attacks on the model always coincided with an increase of collisions in the air and the relatively small common terns were considerably hindered in their

TABLE 45

Number of attacks of different species of breeding birds on a stuffed herring gull placed inside and at 15 m from the edge of the sandwich tern colony

Attacking species of breeding bird	Stuffed gull in colony		Stuffed gull at 15 m from colony-edge	
	Average number of attacks per minute	(range)	Average number of attacks per minute	(range)
Common tern	64	(14-165)	85	(11-161)
Black-headed gull	30	(6- 97)	5	(0- 28)
Sandwich tern	20	(1- 40)	2	(0- 10)
Total	114	(41-208)	92	(31-164)

movements by the black-headed gulls and sandwich terns. It is likely, therefore, that the increase of the number of attacks by common terns was caused by a decrease of the number of attacks of black-headed gulls and sandwich terns.

3.2.4. Effects of the behaviour of common terns, black-headed gulls and sandwich terns upon the predators and its implications for the sandwich tern.

For several reasons it was not possible to study the effect of the reactions of the different breeding birds on the predatory gulls by means of an experiment. Unlike that of black-headed gull predation, a comparison of predation success of herring gulls and black-backed gulls on eggs laid out at different distances from incubating birds promised little success. Firstly, the predatory herring gulls and great black-backed gulls were not very interested in eggs. Secondly, we expected that predatory black-headed gulls would have taken such eggs long before the greater gulls would have discovered them. An experiment with chicks seemed feasible, but ethical reasons kept me from experimentation with living chicks. Thus, the effect of the behaviour of the different breeding birds on the gulls had to be gathered by direct observations on predation under natural circumstances.

The reactions of herring gulls and great black-backed gulls to the attacks of common terns, black-headed gulls and sandwich terns were similar for the three attacking species but varied with the frequency and intensity of the attacks. Furious attacks and chases in the air always caused the predatory gulls to accelerate their speed. If the predators were approached at short distance, they usually made swerving movements to avoid bodily contact. On the ground in the colonies, herring gulls and great black-backed gulls defended themselves against the attacks by pecking upwards. Sometimes the predator repeatedly turned around to face attacking birds approaching from different sides. Actually the attacks of the different breeding birds very seldom resulted in bodily contact with the predator. However, we always got the strong impression that the predators had to exert themselves to the utmost to avoid such contact. This is supported by the fact that the stuffed gull was very often pecked at or hit. Moreover, pecking and hitting a predator under natural circumstances frequently occurred if this predator failed to react appropriately to the attacks of the breeding birds. In 1965 and 1966 a number of sick or wounded herring gulls occasionally stranded in the colonies. Because they were very weak and did not noticeably react to attacks, the breeding birds approached them very closely and often pecked at them. One adult herring gull, landed in a mixed colony of sandwich terns and common terns, was killed in 5 min. by circa 250 pecks of common terns.

These observations demonstrate that the attacks of the different breeding birds can be harmful to a predatory gull if it does not adequately react to it. The fact, that above the colonies herring gulls and great black-backed gulls were forced to fly at great speed, and on the ground, had to defend themselves continually, strongly suggests that the attacks have a considerable influence on the searching efficiency and thus on the predation success of these predators. The following observations support this: (1) Sometimes herring gulls or great black-backed gulls actively hunted on the island for more than an hour before a chick was caught. Usually these gulls roamed an area where hundreds or thousands of chicks were present. Within such an area, a large number of chicks or eggs was always visible from a large distance. If we realise, that herring gulls have a keenly developed eye-sight it is evident that the gulls could have seen the eggs and chicks present in the area. The fact that they often hunted for a long time without being successful can only be explained by distraction and mechanical interference by the continuous attacks of the breeding birds. (2) Herring gulls usually hunted in dense colony areas where a large number of chicks and consequently a large number of adults were present. However, the proportion of successful predation attempts made in dense parts of the colonies was much smaller than in an area where only a small number of birds was present. (3) In parts of the colony where the predatory gulls were attacked most vigorously, the proportion of failed landings was highest. (4) On several occasions I observed that a chick caught by herring gulls or great black-backed gulls was subsequently dropped again. This always happened in a situation in which the predators were vigorously attacked.

From these observations we conclude that the attacks of common terns, black-headed gulls and sandwich terns must be of great protective value for the brood and that the effectiveness of these attacks merely depends on frequency and intensity. It has been argued before that the sandwich tern tends to stay on the nest as long as possible. Probably this behaviour is especially effective in combination with the furious attacks of common terns and black-headed gulls. On a few occasions I witnessed herring gulls taking a prey out of the middle of a large group of sandwich terns. The terns present remained on their nests until the gull swooped down to the ground. At that moment the predator apparently had already fixed upon its prey, which in all cases was the exposed brood in a single black-headed gull's nest situated between the nests of terns.

After the sandwich terns have left the colony area, they spread over a relatively large area. In this stage the adults are much more inclined to fly up, to attack and to chase predators than before. Nevertheless common terns and black-headed gulls play a major role in defending the area. The sandwich terns, therefore, will also importantly benefit from their neighbours in this situation.

On Griend herring gulls and great black-backed gulls are attacked more frequently and more intensively by common terns than by black-headed gulls and predation by herring gulls and great black-backed gulls on eggs and chicks of the common tern was much less than on the brood of the black-headed gull (Table 45). Consequently, we might expect. that the sandwich tern is protected more by neighbouring common terns than by black-headed gulls.

3.3. BEHAVIOUR IN RESPONSE TO OYSTERCATCHERS

On Griend in many places the oystercatcher nests among the sandwich tern. Usually these oystercatchers were ignored by the terns except when they came very close. Sandwich terns have only been recorded to threaten sitting on their nest, when they were approached within a distance of 50 — 150 cm. On a few occasions both species made contact with the bill. However, I never saw a sandwich tern leaving its nest to attack an oystercatcher. In all cases threatening from the nest seemed a very effective defense of the brood against this predator.

The reactions of the oystercatchers to the anti-predator behaviour of the sandwich tern varied from one case to the other. In the neighbourhood of a nest of their own the oystercatchers either avoided, ignored or averted the threatening terns. Sometimes, an oystercatcher's nest was enclosed by a group of sandwich tern nests. Then, the oystercatchers had to run the gauntlet or to fight themselves through to reach their nest. In such a situation nest relief sometimes took more than an hour and very often an enclosed nest was deserted. Oystercatchers penetrating in a part of the sandwich tern colony where they were not nesting, usually reacted to the threat of the terns by immediately withdrawing from the territory of the threatening terns. Since there is a great similarity in the reactions of the sandwich tern to black-headed gull and oystercatcher, it may be assumed that a tern's nest temporarily left is protected by neighbouring nest-owners against predation by oystercatchers in a similar way as it is protected against predation by black-headed gulls (II.3.1.2.).

Common terns and black-headed gulls usually reacted to an oystercatcher as soon as this predator approached their nests within a few meters. The incubating nest-owners first reacted with calls, and if the intruder persisted, would leave the nest and attack it. Usually the owners of one nest only were involved in such attacks. I never saw common terns or black-headed gulls flocking above an oystercatcher or attacking it in great numbers. (It should be noted that the distance at which common terns and black-headed gulls reacted to oystercatchers was relatively large in places where this predator was not present as a breeding bird.)

Depending on whether the predator nested in the neighbourhood or not, attacks were either avoided, ignored or averted, or resulted in a fight. No special observations or experiments have been done to test the possible influence of the anti-predator behaviour of common tern and black-headed gull against oystercatchers on the breeding success of the sandwich tern. However, it may be assumed that sandwich terns nesting within a distance of a few meters from common terns or black-headed gulls are protected by these species against oystercatcher predation, when the nest is temporarily left. This is confirmed by an observation done during the experiment described in II.3.1.4: Six times an oystercatcher attempted to rob an egg placed at a distance of 1 m from a sandwich tern nest and at 2 m from a black-headed gull nest and in all cases the nesting gull prevented predation.

3.4. BEHAVIOUR IN RESPONSE TO OTHER AVIAN PREDATORS

3.4.1. Reactions of common terns, black-headed gulls and sandwich terns.

To a peregrine: The presence of a peregrine almost exclusively evoked flight reactions in all species of breeding birds present on the island. Even if the predator was still at a relatively large distance, common terns, black-headed gulls and sandwich terns reacted with the formation flight: They all flew up from their nests towards the open sea in large flocks. Within each flock the birds synchronously made quick and unpredictable changes in direction.

On 26 April 1967 a peregrine hunting on waders in the neighbourhood of the island was followed for five minutes. Only twice a common tern attacked this predator. Another peregrine, this time not hunting, was present above the island for two minutes on 14 May 1967. This time the predator was pursued and attacked by roughly 50 common terns.

To large raptors, owls and grey herons: The appearance of large raptors (buzzard, honey-buzzard, hen harrier, marsh harrier and osprey), short-eared owl and grey heron evoked flight reactions as well as attack reactions in the different species of breeding birds. Usually the predators were recorded when flying over the island. Their appearance immediately caused all common terns, black-headed gulls and sandwich terns to fly up. Sometimes formation flights were made on a large scale. In all cases the predators were also chased and attacked by a varying number of common terns (up to a hundred) and blackheaded gulls (up to 50).

Ospreys were the only large raptors which alighted near the island. On two occasions the reactions of the different breeding birds to this predator were quantified (Table 46). In both cases the common terns attacked more often than the black-headed gulls and the latter more often than the sandwich terns.

To smaller raptors and crows: A number of species of avian predators including the smaller raptors (hobby, kestrel, merlin and sparrow hawk) and crows (jackdaw, jay and carrion crow) almost exclusively evoked attack-reactions within the different breeding birds. Usually common terns, black-headed gulls and sandwich terns nesting within a relatively small area (a radius of fifty to a hundred meter around the predator) were involved in these attacks. In all cases a large number of common terns (up to a hundred) and black-headed gulls (up to seventy) attacked

TABLE 46

Anti-predator behaviour of Common Terns, Black-headed Gulls and Sandwich Terns in reaction to Osprey and Sparrow Hawk present on the island

Date	Predator species	Average number of attacks per minute made by			Average number of birds in flock		
		C	B	S	C	B	S
15/5-1967	Osprey	14	1	0	25	few	none
20/6-1967	Osprey	101	0	0	125	few	50
7/5-1970	Sparrow Hawk	72	2	0	300	10	none

C = Common Tern, B = Black-headed Gull, S = Sandwich Tern.

and chased the predators (Plate 17). Sometimes the sandwich terns concentrated in large number (up to several hundreds) above them, but only a relatively small number was involved in attacks. The reactions of common terns, black-headed gulls and sandwich terns to a sparrow hawk, are presented in Table 46. It follows that the common terns attacked more often than the black-headed gulls and the latter more often than the sandwich terns.

Summarizing we conclude that to a large number of avian predators including raptors, owls, grey herons and crows, common terns react more aggressively than black-headed gulls, whereas black-headed gulls react more aggressively than sandwich terns. When a predator endangers the adults only, fleeing elements dominate in the reactions of the different species of breeding birds. Elements of approach and attack, however, dominate when the predator is a danger to the brood.

The information presented above gives only a rough survey of the reactions of common terns, black-headed gulls and sandwich terns to the various predator species. The reactions of these breeding birds varied considerably with the behaviour and appearance of the predators. A predator which usually evoked aggression in the breeding birds was sometimes reacted upon with flight if it had not been noticed from a distance and thus appeared very suddenly.

It is also important whether a certain predator appeared in day-time or in the evening. In the evening the tendency of the breeding birds to flee was relatively high, even to a predator which would evoke attack during the day. Moreover, the tendency to flee was relatively high in the beginning and at the end of the breeding season. In the middle of the season, when the eggs hatched, predators were attacked most frequently and most vigorously.

3.4.2. Effects of the behaviour of common tern, black-headed gull and sandwich tern on the predators.

Reactions of the peregrines to the behaviour of the different breeding birds have not been observed. With respect to the black-headed gull this is in agreement with the observations made by MANLEY and KRUUK (KRUUK, 1964).

In flight as well as on the ground, buzzard, honey buzzard, hen harrier, marsh harrier and osprey did not noticeably react to the behaviour of any of the breeding birds present. Grey herons always showed strong reactions to the attacks of both common terns and black-headed gulls. Even relatively weak attacks were reacted upon by marked swerving movements and long swoops. In a few occasions vigorous attacks probably forced the herons to land on the flats around the island. Short-eared owls usually reacted to the close approach of either common terns or black-headed gulls with quick swerving movements.

No reactions to the behaviour of the breeding birds has been recorded in merlins or hobbies passing the island singly. However, kestrels and sparrow hawks — species which often stayed on the island for a relatively long time — always reacted with quick swerving movements when closely approached by common terns and black-headed gulls. On the ground both species often settled in the neighbourhood of patches of high vegetation, thus avoiding the close approach of the attacking breeding birds. Kestrels usually alighted on a high beacon construction where they were safe from any further attack. In the air jackdaws and carrion crows reacted to attacks with a fast swerving flight. Especially the jackdaws had difficulties in avoiding the many attacks of common terns and black-headed gulls. On the ground both jackdaws and carrion crows hid themselves in patches of dense vegetation.

Summarizing we conclude that many predator species showed avoidance reactions in response to the attacks and chases of common terns and black-headed gulls. The anti-predator behaviour of these terns and gulls, therefore, should be regarded as effective in driving off predators from the breeding area.

3.5. BEHAVIOUR IN RESPONSE TO MAN

3.5.1. Reactions of the sandwich terns.

On the nest.

When I approached the sandwich tern colonies on the island the nesting terns usually flew up and immediately alighted again when I was still at a distance of a hundred meters. Thereupon they flew up in mass when I had approached the nearest breeding birds within a distance of 20-50 m and concentrated in a large and noisy flock at 5-15 m high above the colony. Attacks were rarely made. They consisted of faint aerial swoops which never came within a distance of 1-2 m from the predator.

The tern's short up-flight at a hundred meter distance probably was a reaction to the alarm calls and the behaviour of other species nesting in the neighbourhood. Observations from a hide showed that common terns, arctic terns, black-headed gulls and oystercatchers always were the first to react to a man approaching the colony. In turn the terns nearly always reacted immediately to the alarm calls of the other species.

When I did my work in the colonies I got the strong impression that in reaction to human disturbance the terns tended to lure away their chicks from the nest. Human disturbance thus seemed to accelerate nest leaving. In order to study this reaction, in 1970 in three different colonies data have been collected on the time elapsing between hatching and leaving. Between the colonies human disturbance varied as follows:

(1) Colony South Kaap was visited every day. Nests were checked and chicks were ringed.

(2) Colony North Point was visited every day. Nests were checked, but the chicks were not touched.

(3) Colony North II was watched from a hide on a platform at a distance which did not cause any noticeably disturbance. Usually the terns did not fly up even when I entered or left the hide. Nests were checked through a telescope.

The results are given in Table 47. For the 2-egg clutches the time lapse between hatching and leaving is given in relation to the hatching of the first egg. On the average leaving took place 1.2 (1-egg clutches) and 0.8 (2-egg clutches) days later in colony North II than in colony South Kaap. The time lapse between hatching and leaving of 2-egg clutches in colony North Point and colony South Kaap differed only

TABLE 47

The time elapsed between hatching and leaving in three colonies with a different degree of human disturbance

Colony	Hatching period	Method of checking nests	Number of nests checked	Average number of days between hatching and leaving (number of observations between brackets)	
				1-egg clutches	2-egg clutches
South Kaap	8/6-13/6	daily visits (chicks ringed)	55	3.7 (13)	4.7 (42)
North Point	7/6-18/6	daily visits	34	4.5 (2)	4.8 (32)
North II	19/6-22/6	observations from hide	53	4.9 (22)	5.5 (31)

slightly. In the 1-egg clutches leaving appeared to be later in colony North Point than in colony South 'Kaap'. However, only two 1-egg clutches were followed in colony North Point and the accuracy of the method is ± 1 day (daily checks). The small number of nests studied in this colony thus might have caused the difference observed.

Summarizing we may conclude that the time lapse between hatching and leaving decreases with roughly one day if a colony is visited daily. It remains uncertain whether or not ringing, in addition to walking through the colony, has an effect on nest-leaving.

After nest-departure.

In reaction to a man approaching the area, sandwich terns which had left the colony usually reacted with flying up when the intruder was still at a distance of a hundred meters. Thereupon they concentrated in a sparse flock above the predator. In general the intensity and the frequency of attacks was very low, although much higher than in the colony situation. Attacks consisted of aerial swoops which rarely came within a distance of a meter from the predator. Actual contact with the predator was not recorded. Usually only one or a few birds were involved in attacks at the same time. In general, the tendency of a tern to attack appeared to be relatively high in an area where the density of chicks, and consequently the number of adults present, was low. Thus, attacking the predator apparently was inhibited by a high density of chicks or adults.

To Man, chicks may react in two different ways (Plate 10 and 11): (1) They may crouch and keep motionless, preferably in dense vegetation. (2) Chick of over two weeks old may also concentrate in dense groups moving away from the predator. Groups of chicks counted up to many hundreds of individuals. They clearly preferred to stay on the island. However, if they were closely approached, they might enter the tidal flats or, at high tide, swim away from the predator.

After the chicks have left the nest they spread over the vegetated part of the island. To assess breeding success, parts of the island were searched for chicks in the post hatching period. In an area which had not been searched before, the majority of chicks was easily found. They usually hid under a patch of vegetation where they had made funnel-shaped shelters, often with conspicuous white droppings at the entrance. The appearance of these shelters indicated that these chicks had remained for rather a long time on exactly the same place. However, immediately after I had searched a particular area, the chicks present always moved to another part of the island.

3.5.2. Effects of the behaviour of the sandwich tern upon the predator.

The reactions of the adult terns very likely have no effect on a person, collecting eggs or chicks. Eggs and chicks have a cryptic colouration and small groups of nests may be difficult to find. However, once the first egg of a group is discovered it is very easy to find the other eggs. If the behaviour of the adults have a distractive effect, it should be regarded to be very small.

The probability of a chick being discovered strongly decreases after nest-leaving and the tendency to accelerate nest-leaving in reaction to human disturbance seems very effective. Nevertheless, searching the island in the post hatching period, I could recover some 50% of the number of chicks present in a particular area. This relatively high figure might be partly caused by the training I had, for unexperienced people usually did not find half as much. Early leaving also has a disadvantage: it increases the chance of predation by black-headed gulls and, with respect to the second-hatched chicks of two-egg clutches, to straggle behind.

The tendency of the chicks to move from a particular area after considerable disturbance (as for instance during my presence in the colonies) has the obvious effect that a predator will be less succesful if he returns to the same area. Moreover, if he discovers where the chicks have gone, the chicks are still less vulnerable because the fresh hiding places are not marked with a patch of white droppings. It is understandable that the chicks are not constantly on the move; this might make it considerably more difficult for the parents to feed them and it would be a disadvantage for rapidly finding hiding places.

Considering the high proportion of chicks that was recovered by searching the area, the tendency of the older chicks to concentrate in groups walking away from the predator seems another very effective reaction. One has to walk over a long distance and to run fast to catch only a few chicks. Without special aids (II.1.2.) I was not very successful in my effort to catch chicks from these groups.

3.5.3. Reactions of common terns and black-headed gulls and their effects.

On Griend, the reactions to Man of common tern and black-headed gull have not been given special attention. Detailed information on the anti-predator behaviour to Man of the black-headed gull is given by KRUUK (1964). He mentions that the time of the day, the phase of breeding and the behaviour of the predator are important factors which determine the reactions of the gulls. In the middle of the breeding season, a person walking or standing in the colony usually causes a large number of birds to fly up, forming a flock which may consist of hundreds of birds. The gulls attack by aerial swoops and often defecate at the lowest point of a swoop. The frequency of attacks may be high and the gulls may come very close; about hatching time they often hit the intruder.

On Griend it was evident that in response to Man, the black-headed gull reacted more intensively and more frequently with attacks, than the sandwich tern. Besides, the response area of the black-headed gull was much greater than in the sandwich tern. However, in all respects the reactions of the common tern were even stronger than in the black-headed gull. In the common tern colony about hatching time, a human intruder was almost constantly attacked by several individuals sometimes approaching from different sides. Very often the terns repeatedly pecked at the intruder and defecated at the lowest point of the swoop.

Considering the possible effect of the attacks of black-headed gulls and common terns on Man, it seems likely that determined egg-collectors will not take one egg less because of the behaviour of these breeding birds. However, if one has not taken precautions against being hit with the legs, pecked with the bills and bombarded with droppings, one will soon feel uncomfortable. In the middle of a common tern colony, in the hatching period, it takes constant attention to avoid the persistent attacks. Therefore, the attacks of black-headed gull and common tern might lessen Man's searching efficiency and the camouflaged brood of sandwich terns nesting among black-headed gulls and common terns might gain advantage of it.

3.6. BEHAVIOUR IN RESPONSE TO OTHER MAMMALIAN GROUND PREDATORS

A variety of mammalian ground predators including polecat, stoat, weasel, hedgehog and rat have been recorded in sandwich tern colonies in the Netherlands. (Table 33). On Griend these predators do not occur. However, for a good understanding of the anti-predator behaviour of the sandwich tern and the possible significance of associative nesting with black-headed gulls and common terns, it seemed necessary to gain some insight in the reactions of the terns and the gulls to this group of predators. Therefore, two experiments have been done with a stuffed and a living mammalian predator in analogy to the experiments with a stuffed herring gull as described in II.3.2.3.3. and II.3.2.3.4. The aim of these experiments was (1) to study anti-predator behaviour of sandwich tern, common tern and black-headed gull and (2) to investigate if, and to what extent, common terns and black-headed gulls participate in attacking mammalian ground predators present in the sandwich tern colony.

3.6.1. Experiments with stuffed and living models

3.6.1.1. *Material and methods.*

A stuffed polecat and a living brown-coloured laboratory rat were used in the experiments. The rat was placed in a circular roofless wire-netting cage, measuring one meter in diameter and 50 cm high. Within this cage the animal could move freely. Control observations showed that the breeding birds did not noticeably react to the cage alone. The presentation of the predators into the colony and the method of observation was similar as described for the stuffed herring gull (II.3.2.3.2.).

3.6.1.2. *Analysis of the reactions of common terns, black-headed gulls and sandwich terns.*

In 1968 an experiment was done to study the reactions of common tern, black-headed gull and sandwich tern to a stuffed polecat and a living rat placed inside the colony and at 5 and 15 m from the colony. edge. As it was our aim to study the reactions of each species separately, relatively unmixed parts of the colony were selected for the experiment. At each location chosen one series of 5-min. tests was done. Tests within one series were done with intervals of a day.

Description of the reactions:

(a) Stuffed polecat.

If the polecat was placed inside the colony, a varying number of sandwich terns either concentrated in a dense flock above the predator or flew around in wide circles. The majority of terns nesting at a distance of more than 150 cm from the model usually alighted soon. However, during a five-minute period of observation the terns might fly up in mass repeatedly, flying around in a way which resembled a weak formation-flight. Attacks on the model were small in number, relatively weak and never approached the predator within a distance of a meter. Usually the sandwich terns did not noticeably react to the polecat placed at 5 and 15 meter from the nests.

Depending on the distance of the model to the nearest nests, a varying number of common terns or black-headed gulls concentrated above the predator. In the course of each inside-colony test the majority of birds nesting at more than two meter from the model alighted on their nests. In the outside-colony tests nest-owners usually remained on the nest all the time long. The attacks of both common terns and black-headed gulls consisted of aerial swoops of varying intensity. Inside the colony the common terns sometimes pecked the model. Black-headed gulls often approached the model within a distance of 50 cm. They usually struck at it with their feet, but actual contact was never recorded.

(b) Living rat

Inside the colony the reactions of the different breeding birds to a living rat closely resembled the reactions to the stuffed polecat. In general the number of birds concentrated in a flock, as well as the number of attacks made, were smaller. Moreover, the attacks were less intensive and the rat was never touched.

We often got the impression that the presence of the cage prevented the breeding birds from closely approaching the rat. In a few cases the rat climbed out of its roofless cage, which made it possible to study the reactions of the breeding birds in a more natural situation. It was shown, that the intensity of the attacks of common terns and black-headed gulls strongly increased if the predator approached a nest. However, even within a distance of 50 cm from a nest neither common terns nor black-headed gulls were recorded to touch the predator.

Besides intensity-differences, the reactions of the different breeding birds to polecat and rat mainly differed from the reactions to a stuffed

Plate 15. A black-headed gull making an attempt to seize the fish of a sandwich tern which has landed near its nest. The neighbouring terns threaten from their nests by pointing their bills into the direction of the gull and by producing 'gackering' calls.

Plate 16. A common tern attacking a stuffed herring gull placed at some distance from its nest.

15

16

17

18

herring gull in the way in which the birds concentrated in a flock. A flock formed in reaction to polecat and rat usually was very tight and rather close to the ground. Within this flock individual birds did not fly around, but kept position by hovering. The formation of this so-called 'ground-predator flock' was strongest in the common tern. It was less pronounced in the sandwich tern and often did not occur at all in the black-headed gull. Observations on a rat which freely moved through the colonies showed that a 'ground-predator flock' of common terns and black-headed gulls follows the predator very precisely through the area. If the predator was difficult to see, for instance due to darkness or dense vegetation, the tendency to form a 'ground-predator flock' was very strong. In such a situation attacks were hardly recorded.

Attack-frequency.

Figure 35 gives the frequency of attacks of common terns, black-headed gulls and sandwich terns on the living rat and the stuffed polecat placed at different distances from the nests. It shows that for each species of breeding bird the attack-frequency strongly decreased with increase of the distance between the predator and the nests.

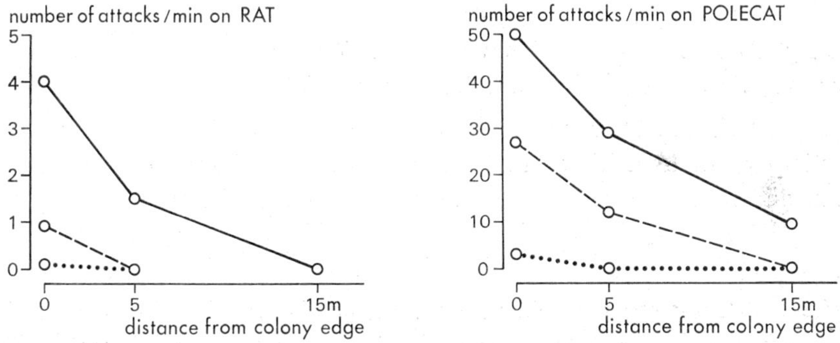

Fig. 35. Average number of attacks per minute made by common terns (solid line), black-headed gulls (broken line) and sandwich terns (dotted line) on a living rat (left) and a stuffed polecat (right) placed at different distances from the nests.

Table 48 gives the size of the area defended, expressed by the distance at which the breeding birds attacked the predators half as much as they did near their nests. To rat and to polecat the size of this area was greatest in the common tern, smaller in the black-headed gulls and smallest in the sandwich tern.

Plate 17. If a predator appears on the island, common and arctic terns react more agressively than black-headed gulls and the latter more than the sandwich terns. The picture shows a carrion crow being chased and attacked by common terns.

Plate 18. House-hunting sandwich terns inspecting a potential nest site at a short distance from incubating conspecifics. The house-hunters are threatened by the incubating birds.

TABLE 48

Area defended by common tern, black-headed gull and sandwich tern against a living rat and a stuffed polecat

Species of breeding bird	Radius in meters of the area defended against	
	Living rat	Stuffed polecat
Common tern	4	6½
Black-headed gull	<1	2½
Sandwich tern	$\ll 1$	<1

In Figure 35 the frequency of attacks on both predators was relatively high in the common tern colony, intermediate in the black-headed gull colony and relatively small in the sandwich tern colony. As this might be a consequence of the varying number of breeding birds present within the colonies of the different species, the ratio of the number of attacks and the number of birds concentrated above the predators is given in Table 49. Only inside-colony tests are included in the calculation. It follows, that the 'ıelative attack-frequency' on the rat as well as on the polecat decreased in the order common tern — black-headed gull — sandwich tern.

3.6.1.3. *Combined attacks of common terns, black-headed gulls and sandwich terns on rat and polecat presented in the sandwich tern colony.*

To test if, and to what extent, common terns and black-headed gulls participate in attacking mammalian ground predators present in the sandwich tern colony, an experiment has been done between 27 May and

TABLE 49

Relative attack-frequency of common tern, black-headed gull and sandwich tern on rat and polecat presented inside the colonies of the different species

Species of breeding bird	Average number of attacks per bird per minute on:	
	Living rat	Stuffed polecat
Common tern	0.46	0.86
Black-headed gull	0.08	0.90
Sandwich tern	0.01	0.06

6 June 1967 by placing the living rat and the stuffed polecat inside and at 15 meter from the edge of the sandwich tern colony. Tests were done at eight different locations randomly chosen from the sandwich tern colony. Each location was used once for an inside-colony test and once for a test at 15 meter from the edge. Tests inside and outside the colony were done with intervals of two days. Each test lasted five minutes. The number of common terns and black-headed gulls nesting among the sandwich terns varied from one place to the other.

Table 50 gives the total number of attacks released by the predators in the eight 5-min. observations, together with the proportion of these attacks made by common terns, black-headed gulls and sandwich terns. It follows that the predators inside as well as at 15 m from the edge of the colony were attacked much more often by common terns and black-headed gulls than by the sandwich terns.

TABLE 50

Proportional contribution of different species of breeding birds to the attacks on a living rat and a stuffed polecat placed inside and at 15 m from the edge of the sandwich tern colony

Species of breeding bird	Living rat		Stuffed polecat	
	Inside colony	15 m from colony-edge	Inside colony	15 m from colony-edge
Common tern	74.8%	100.0%	54.4%	98.6%
Black-headed gull	16.0%	0.0%	32.0%	1.0%
Sandwich tern	9.2%	0.0%	13.6%	0.4%
Total number of attacks	206	21	670	225

From the above results we conclude that the frequency of attacks on and the area defended against a living rat and a stuffed polecat was greatest in the common tern, smaller in the black-headed gull and smaller again in the sandwich tern. Moreover, it can be concluded that the rat and the polecat presented inside and at 15 m from the edge of the sandwich tern colony were attacked more often by common terns and black-headed gulls than by the sandwich terns itself.

3.6.2. Effects of the behaviour of common tern and black-headed gull on mammalian ground predators and its implications for the sandwich tern.

If we accept the above results applicable to the natural situation, we can expect that in case of predation by a rat or a polecat, the sandwich tern will benefit from neighbouring common terns and black-headed gulls, provided that (1) the behaviour of common terns and black-headed gulls influences predation success of the predators concerned and (2) that the sandwich terns nest within 'predator-reaction-distance' of common terns and black-headed gulls.

Unfortunately, it was not possible to investigate the effect of the anti-predator behaviour of terns and gulls on mammalian ground predators on Griend. However, information is given by KRUUK (1964), who studied the effect of anti-predator behaviour of black-headed gulls to stoat and hedgehog. His observations showed that stoats strongly react to the attacks of black-headed gulls and usually avoid black-headed gull colonies. With respect to the hedgehog it was concluded, that the gulls' attacks provide good protection, although this predator managed to eat many broods. With respect to the effectiveness of the anti-predator behaviour of the common tern only one observation is available. ROOTH (pers. comm.) saw a polecat which was furiously attacked by common terns. In reaction to these attacks the predator ran away at high speed and soon disappeared from the area. This information, although very incomplete, leads to the conclusion that common terns and black-headed gulls probably defend their brood effectively against the majority of species of mammalian ground predators recorded in colonies in the Netherlands (Table 33).

In the Griend colonies roughly ninety percent of the sandwich terns nest within a distance of four meters from a common tern or a black-headed gull. The radius of the area defended with attack to a stuffed polecat, stoat and hedgehog and to a living rat varies from 4 to 6½ meters in the common tern and from less than 1 to 2½ meters in the black-headed gull (Table 48 and KRUUK, 1964). As it is not likely that in general inter-nest distances between sandwich terns and common terns and black-headed gulls are much different from the Griend situation, we can expect that many sandwich terns benefit from the anti-predator behaviour of neighbouring common terns and black-headed gulls to mammalian ground predators.

Besides attacking a predator, common terns and black-headed gulls may react already with flocking when a predator is still at a relatively large distance. To a stoat for instance, black-headed gulls react with flocking at several hundred meters from the colony (KRUUK, 1964). The effect of flocking on the behaviour of a mammalian ground predator and its predation success is still poorly understood. If flocking acts as a warning or if it distracts or deters the predator, the anti-predator behaviour of common terns and black-headed gulls might even contribute to the survival of the brood of the sandwich tern over a much greater distance than it is suggested by the frequency of attacks on the stuffed predator-models.

With respect to the chance to come in contact with mammalian ground-predators, the location of sandwich tern colonies relative to neighbouring black-headed gulls and common terns is a point of special interest. Several authors (SALOMONSEN, 1947; LIND, 1963) have mentioned that sandwich terns tend to settle in the middle of an already established black-headed gull colony. In the Netherlands this type of colony is also known from a number of breeding places. The distribution pattern of the colonies of the sandwich tern, the common tern and the black-headed gull may also have a more or less fixed zonation. In the province of Zealand (in the southern part of the Netherlands), before the Deltaplan was executed, sandwich terns, common terns and black-headed gulls usually nested in salt marshes at the seaside of embanked land. Within these areas the sandwich terns always nested close to the waterside, whereas common terns and black-headed gulls had their nests at the more inland side of the sandwich terns (LEBRET, pers. comm.). In both cases the distribution pattern of the colonies implicate that mammalian ground predators have to pass common terns or black-headed gulls on their way to the sandwich tern colony. It is likely, that the predators usually will be satiated before having reached the sandwich tern colony. Moreover, the sandwich terns will have a maximal profit of the anti-predator behaviour of the common terns and the black-headed gulls present in the neighbourhood.

3.7. CONCLUSIONS

The anti-predator behaviour of the sandwich tern strongly depends on the species of predator to which it is directed and on the stage of breeding of the terns. To a predator which endangers the brood only, sandwich terns react with sitting down on the nest, threatening from sitting position. To a predator which is a danger to the brood and to the adults, the terns react with flying up; sometimes they attack the predator, but usually the attack frequency and intensity is low. In reaction to a predator which endangers only the adults, flying up from the nest followed by escape behaviour is shown. The tendency of the terns to fly up and to attack a predator strongly increases after they have left the nest, shortly after hatching.

Throughout the whole breeding season, common tern and black-headed gull tend to fly up in reaction to nearly all predator species. Predators on the brood are vigorously attacked.

Observations under natural circumstances and experiments with predator models have shown that on a variety of predator species (a) the total number of attacks made, (b) the size of the area defended, (c) the number of attacks per individual and (d) the closeness of attacks, is greatest in the common tern, smaller in the black-headed gull and smaller again in the sandwich tern.

The anti-predator behaviour of the sandwich tern defends the brood effectively against black-headed gulls, oystercatchers, herring gulls and great black-backed gulls. The behaviour of the terns seems to give only little protection to the brood when predators are involved which endanger both adults and brood. The brood of the sandwich tern is highly vulnerable to predation by mammalian ground predators.

The behaviour of predators reacting to the attacks of common tern, black-headed gull and sandwich tern mainly depends on attack frequency and intensity. Aerial predators are successfully driven off from the nesting area by the combined attacks of the different breeding birds. The attacks are probably also very effective against a number of mammalian ground predators, but little information is available.

There is no doubt that in general the sandwich tern gains protection against predators from common terns and black-headed gulls nesting in the neighbourhood. This protective influence has been demonstrated in case of predation by black-headed gulls, herring gulls and great black-backed gulls. In case of predation by herring gulls and great black-backed gulls it has been shown that the tendency of the sandwich tern to stay on the nest as long as possible, is especially effective in combination with the tendency to fly up shown by common tern and black-headed gull.

4. DISCUSSION

It has been argued in chapter II.1 that a number of aspects of nest distribution, *e.g.* nest density, colony size, group synchronization and island synchronization, contribute to reproductive success in the sandwich tern. The data on mortality agents showed, that predation was one of the most important causes of egg and chick losses. In Chapter II.2 and II.3 predators and anti-predator behaviour have been studied. In this chapter we shall discuss in succession: (1) The mechanisms by which the various aspects of nest distribution studied might reduce predation, (2) the advantages and disadvantages of associative nesting with another species and (3) the selective advantage of the sandwich tern nest distribution pattern with respect to some environmental factors other than predation.

4.1. THE MECHANISMS BY WHICH THE DISTRIBUTION OF THE NESTS IN SPACE AND TIME MIGHT REDUCE PREDATION

Nest density.

Our data show that the majority of sandwich terns nested at a distance of 31-40 cm from the nearest neighbour. These terns were more successful in hatching their eggs than pairs nesting at both higher and lower densities. Egg-mortality in nests with a relatively small minimum inter-nest distance mainly referred to addled eggs, whereas predation was the most important mortality factor in nests with a relatively large minimum inter-nest distance. This suggests that in the sandwich tern, predation selects against dispersed nesting.

On Griend, black-headed gulls and oystercatchers are the most important egg-predators. Against these predators the sandwich tern defends its brood by threatening from the nest. This method of defense protects the brood very effectively.

Observations under natural circumstances and the 'egg-line' experiment (II.3.1.2.) have shown that a clutch temporarily left by the parents gains protection from nearby neighbours. The protective value of nearby breeding birds increases with decrease of distance. This explains the positive correlation between nest density and egg-predation.

Colony-size.

Pairs nesting in colonies of over 200 nests were more successful than birds nesting in smaller colonies. This mainly referred to a difference in the proportion of eggs taken by predators. Differences in predation on eggs between small and large colonies would only be understandable if the predator-prey ratio is different, or if the eggs in the smaller colonies are more often exposed to predation. The most important egg-predators on Griend, the black-headed gull and the oystercatcher, are extremely mobile and regularly visit each breeding concentration of sandwich terns present on the island. Exact data on the predator-prey ratio in colonies of different size are not available, but marked differences certainly did not occur. Also, we never noticed differences between sandwich terns from large and small colonies in their reactions to robbing black-headed gulls and oystercatchers. In contrast, marked differences were observed in the reaction to the presence of humans; in large colonies the breeding birds flew up much later and laighted much earlier than in smaller colonies. In response to herring gulls and great black-backed gulls the breeding birds of the larger colonies rarely flew up. In the smaller colonies, however, they often did so. It has been shown in II.3.2.1. that these differences in behaviour caused large and small colonies to be differentially preyed upon by herring gulls. But this cannot explain the difference in hatching success between small and large colonies on Griend, since on this island the herring gull only takes a small part of the total number of eggs predated. However, we know that black-headed gulls and oystercatchers which have specialised in egg-predation, tend to keep a watch on the terneries, quickly grabbing an egg when the opportunity arises. Several times we noticed that eggs were robbed when the terns flew up in reaction to humans, herring gulls and birds of prey. It is likely, therefore, that egg-predation is different between large and small colonies mainly because in the latter eggs are more frequently exposed to predation (Table 40).

Group synchronization.

Synchronization of egg-laying within a group means that neighbouring nests are present from the beginning until the end of the nesting period. Group synchronization thus contributes to nest density in time and, as a consequence, reduces predation on eggs. However, our data showed that close synchronization is still correlated with a low predation risk if the density aspect referred to above is excluded. This might be explained in two ways:

First, the less synchronized 'tail breeders' of a group always nested later than the highly synchronized 'peak breeders'. Our data show that there is a rapid increase throughout the season of the number of black-headed gulls preying on eggs. As a consequence the predator-prey ratio must have been relatively unfavourable for the tail breeders.

Second, the tail breeders of a group were much more inclined than the peak breeders to leave the nest in reaction to human disturbance or in reaction to the presence of other predators. This may have had the effect that clutches of tail breeders were more often exposed to predation, and — in combination with a relatively high predation pressure — this might explain the relatively high loss of eggs in the tail breeders.

Island synchronization.

Synchronization of egg-laying in the island population enables the sandwich tern to nest in large colonies and this was shown to contribute to reproductive success though reducing the number of eggs robbed. However, island synchronization was also shown to be correlated with hatching success apart from colony size and besides, it was found to be strongly correlated with fledging succes.

As a consequence of island synchronization prey items are only present in large numbers for a relatively short period. If the number of predators within the area is limited, this could be an important mechanism for reducing the number of eggs and chicks taken. On Griend the number of egg- and chick-predators is limited, but rapidly increases in the course of the season. In comparison to the peak breeders the tail breeders of the population have to cope with a relatively large number of predators, whereas their own number is relatively small. As a consequence predation pressure on both groups is greatly different. This may partly explain the difference in hatching and fledging success.

Predation on chicks which have already left the nest is mainly practised by herring gulls and great black-backed gulls. Only a few individuals specialized in preying upon the sandwich tern. Moreover, on Griend the number of gulls actively involved in predation on the sandwich tern was limited by intra specific aggression: usually only one herring gull hunted on the island at the same time, chasing away from its feeding territory all other large gulls. This single herring gull only robbed a very small proportion of the chicks from the peak breeders but among the tail breeders this proportion was much higher; among the birds laying their eggs in June and July in some years it went up as far as 100%.

Cryptic colouration.

It might be suggested that the high nest density of the sandwich tern has a negative influence on the camouflage of eggs and chicks. The way in which the terns defend their brood against predators seems related to this: As long as eggs and chicks are present within the colony, the adults tend to stay on the nest, thus avoiding exposure of the brood. After nest leaving, when the chicks hide in dense vegetation, they change their behaviour and usually react to a predator with flying up and attacking.

With respect to the nesting situation of the sandwich tern CROZE (1970) has drawn attention to the fact that predators tend to develop a searching image for camouflaged prey and that prey species markedly different in appearance from the image object usually are not included in the predators' searching image. CROZE, working in the neighbourhood of the black-headed gull-sandwich tern colony in Ravenglass (Cumberland, U.K.) demonstrated that carrion crows which are predators on eggs of colony nesting Larids, easily develop a searching image. In Ravenglass a small sandwich tern colony is situated in the middle of a large black-headed gull colony. In contrast to the black-headed gull, the sandwich tern defecates near the nest, thus 'painting' a conspicuous white colony area against which the predominantly black and white egg is very well camouflaged. CROZE decided to '. . . .artificially replicate the natural sandwich tern-black-headed gull situation, to see if a unique camouflage pattern on a special blackground in the midst of a frequently preyed-upon area would escape predation (p. 74)'. An experiment was done in which survival of meat rewards hidden under red mussel shells laid out on a pebble beach was compared with the survival of the same reward hidden under black and white painted shells placed in a white painted 'artificial ternery' on the same beach. The wild crows which visited the area had developed a 'searching image' for the red mussels before the 'tern shells' were introduced. It followed from the experiment that the black and white painted 'tern shells' escaped the crow's notice for several days. Besides, it was noticed that the crows were afraid to enter the white painted area. Extrapolating these results to the natural situation CROZE suggests that the white sandwich tern colony, seen from a distance, might be frightening to some avian predators. He suggests that the function of the white colouration is two-fold: 'to camouflage the eggs and chicks and to avoid the most probable searching image in the area, namely that for the eggs of the host species (p. 80)'.

In my opinion several objections can be made against extrapolation of the experimental results to the natural situation. For instance, possible differences in the behaviour of crows to a white washed ternery with and without sandwich terns present are not taken into account. Even if we put aside the various uncontrolled variables which may have influenced the results of the experiment, CROZE's hypothesis is only sound, if in general (1) the predators of the sandwich tern (which in the Netherlands seem to be carrion crows only rarely, see Table 22) hunt by searching image, (2) the background of the ternery contrasts with the environment, (3) the predators react to this contrast in a similar way as do carrion crows and (4) the predators will more easily develop a searching image for the brood of the black-headed gull than for the eggs and the young of the sandwich tern; *i.e.* when these gulls highly outnumber the terns.

On Griend I got the strong impression, that black-headed gulls and herring gulls both hunted by searching image, in the sense that eggs and chicks were generally ignored and only entered the diet of certain individuals after having gained experience with this source of food. However, it always appeared to me that the predators concerned developed a searching image for the brood of the sandwich tern more easily than for the brood of the black-geaded gull. The 'painted' background of the sandwich tern colonies on Griend often did not contrast very much with the environment. Moreover, the various predators on the island never showed fear for the 'white painted' area. Finally, in the Griend situation as well as in other colonies in the Netherlands the black-headed gull does not markedly outnumber the sandwich tern. (With respect to the black-headed gull/sandwich tern ratio, the present situation in the Netherlands is importantly influenced by a recent decline of the number of sandwich terns as a consequence of pollution by organochlorine pesticides (KOEMAN *et al.*, 1967; VEEN, 1967).

Considering the many differences between Ravenglass and Griend, it is uncertain whether or not nesting in association of sandwich terns and black-headed gulls is advantageous to the sandwich tern in the sense of CROZE's hypothesis.

Habitat selection.

The nest distribution of the sandwich tern, in combination with suitable anti-predator reactions, effectively protects the brood from aerial predators. However, to mammalian ground predators, the species (inclusive of the adults) is highly vulnerable. Therefore, the safety provided by the nest distribution pattern of the sandwich tern is strongly dependent on the selection of a habitat relatively free from mammalian ground predators. The behaviour of the terns seems to include special adaptations to avoid such predators as much as possible.

First, sandwich terns tend to select a nesting habitat which for mammalian ground predators is difficult to reach. All sandwich tern colonies known from the Netherlands are situated either on remote and isolated islets or in coastal salt marshes cut off from the mainland by channels. Usually, the nesting area of the sandwich tern is flooded by high tides in spring and winter. Mammalian ground predators, therefore, have to immigrate into the area each spring and summer. By doing so, they often have to walk over a distance of many kilometers, traversing a relatively wet area. Moreover, within the area the likelihood to find prey other than the brood of the terns and the gulls is small, at least in the beginning of the season. (Possibly an exception should be made with respect to the rat, which may feed on plants and carrion.) Considering the many difficulties the predators have to overcome, together with the fact that the predator population-density will be relatively low at the end of winter, the number of predators present in the colony area can be expected to increase only very slowly throughout the season. As the sandwich tern tends to synchronize egg-laying early in the season, the number of mammalian ground predators they have to cope with will be relatively small.

Second, when mammalian ground predators do intrude within the breeding area, the sandwich terns immediately react with avoidance behaviour. In the pre-laying period, in reaction to human disturbance, they may easily move to another place. They probably react similarly to the presence of other ground mammals. The tendency to leave the area after it has been intruded by mammalian ground predators remains throughout the breeding season: colonies with newly laid eggs are easily deserted; nest-departure of small chicks may be speeded up and large chicks may walk a long distance.

Third, there are indications that sandwich terns do not return to a colony area in which they have been preyed upon by mammalian ground predators the year before.

4.2. ASSOCIATIVE NESTING WITH OTHER SPECIES

4.2.1. The Griend situation: advantage or disadvantage.

The following account deals with the question to what extent advantage and disadvantage of associateve nesting with black-headed gulls and common terns balance each other. To answer it, an assessment has to be made of the number of sandwich tern eggs and chicks that would be lost to predators if common terns and black-headed gulls had not been present in the area. We can only make an estimate as far as predation by black-headed gulls, herring gulls and great black-backed gulls is concerned. The possible influence of predators not active in the various years we were observing the terns cannot be incorporated.

In a number of sandwich tern colonies in the Netherlands (Rottumerplaat and Rottumeroog) and in Germany (Knechtsand) the sandwich tern is not preyed upon by black-headed gulls. These colonies are exactly the ones where black-headed gulls are not present as breeders. Thus, predation by black-headed gulls on the brood of the sandwich tern seems a consequence of the presence of these gulls as breeding birds. This view is supported by the fact that the first encounter between a black-headed gull and the brood of the sandwich tern is incidental and that thereupon the gull may develop into an experienced egg-predator.

In different years of observation, the herring gulls and great black-backed gulls on Griend mainly robbed chicks of black-headed gulls and sandwich terns (Table 29). If the black-headed gulls had not been present on the island, only sandwich tern chicks would have been taken by these predators. If the total number of chicks robbed would not change in such a situation, 285 and 124 sandwich tern chicks instead of 174 and 46 would have been taken in 1970 and 1971 (Table 31). This means that the number of sandwich tern chicks fledged would have been 32 more (or 1.2% of the number of eggs laid) in 1970 and 76 less (or 2.1% of the number of eggs laid) in 1971 than was actually recorded in the presence of black-headed gulls in both years (for methods of calculation see II.2.8.).

If the results from both years are considered together, it seems as if advantage and disadvantage of associated nesting roughly balance each other. However, the above calculation only refers to an increase of the risk to be preyed upon by herring gulls and great black-backed gulls as a consequence of a decrease of the total number of eggs and chicks present on the island. We tacitly assumed the number of predatory herring gulls and great black-backed gulls, as well as the number of chicks robbed per gull, to remain constant. This however, is not very likely. It has been demonstrated that the attacks by black-headed gulls on predatory herring gulls and great black-backed gulls are very effective. Therefore, we might expect an increase of predation success of herring gulls and great black-backed gulls if the black-headed gulls disappear from the breeding area. Moreover, the number of gulls involved in predation might increase. Unfortunately I cannot give the influence of attacks of black-headed gulls on the predation success of herring gulls and great black-backed gulls and on the number of these predators present in figures. However, it has been shown that the sandwich terns on Griend greatly benefit from the anti-predator behaviour of the black-headed gulls. Moreover, various authors (DIRCKSEN, 1932; LIND, 1963; ROOTH, 1965) have mentioned that sandwich tern colonies which are not protected by black-headed gulls usually suffer heavily from herring gull predation. These facts strongly suggest that the number of chicks taken from the Griend colonies by herring gulls and great black-backed gulls would have increased considerably if the black-headed gulls had disappeared from the area. Therefore, it is very likely that in the Griend situation the advantage of nesting together with black-headed gulls outweighs its disadvantages.

4.2.2. Associative nesting in other sandwich tern colonies.

The above argument only refers to the Griend situation, which can be regarded as representative for the sandwich tern in general as far as predation by avian predators in concerned. However, it has been shown (II.2.6. and Table 33) that mammalian ground predators have been recorded in a number of sandwich tern colonies in the Netherlands. If predation by ground mammals is a potential threat to the terns, associative nesting with common terns and black-headed gulls might have the following advantages:

(1) Colonies of black-headed gulls and common terns often occur for many consecutive years within the same area. This phenomenon is mainly the result of two behaviour mechanisms: (a) Birds breeding for the first time usually return to the colony of birth and (b) experienced breeders tend to return to the same place were they have bred successfully before. Both mechanisms will lead to a selection of breeding places which are most favourable with respect to the influence of factors determining breeding success *e.g.* food, high water and predation. Within the limits set by the other factors mentioned, predation can be expected to be of great importance in determining whether a particular place is

favourable or not. Indeed several traditional colonies of both black-headed gulls and common terns are situated on small islets or in marshes known to be inaccessible for mammalian ground predators. In contrast with black-headed gulls and common terns, sandwich terns easily occupy an entirely new area for breeding from one year to the other. This behaviour entails that sandwich terns are often unfamiliar with the occurrence and distribution of predators within their new breeding area. The presence of a colony of black-headed gulls or common terns might be a good indication of sites relatively safe from mammalian ground predators.

(2) It has been argued in the case of predation by herring gulls and great black-backed gulls on Griend, that an increase of the total number of terns and gulls present will lower the risk for each individual to fall victim to predation. This advantage of associative nesting will be maximal if the predator population is prevented from adjusting itself to the number of terns and gulls present in the area, which seems to be the case with predation by ground mammals.

(3) Finally, the sandwich tern might greatly benefit from the attacks of common terns and black-headed gulls on mammalian ground-predators (II.3.6.2.).

Summarizing we conclude that in colonies where the sandwich tern has to cope with mammalian ground predators, associative nesting has a number of advantages in addition to the ones already mentioned in the Griend situation. It has been concluded before (II.2.7.) that the Griend situation can be regarded as representative for the situation in sandwich tern colonies in general as far as predation by black-headed gulls and other avian predators is concerned. As it is likely that in the Griend situation the advantages of associative nesting outweigh its disadvantages, it seems justified to expand this conclusion by stating that associative nesting with common terns and black-headed gulls is advantageous for the sandwich tern in general.

4.2.3. Other aspects of associative nesting.

The present study mainly deals with the effect of associative nesting on predation. However, a number of aspects of associative nesting, such as competition for space and food deserve our attention.

Space-competition.

The nesting habitats of sandwich tern, common tern and black-headed gull overlap. Moreover, all species mentioned prefer relatively high places where they are not in danger of being washed away by high tides. However, available space safe against such tides is usually very restricted and as common terns and black-headed gulls occupy their nesting territory much earlier than the sandwich terns, the latter may be at a disadvantage if space competition occurs.

In the Griend colonies, in the pre-laying period, sandwich terns often quarrelled with common terns and black-headed gulls. Usually the common terns and the black-headed gulls finally were supplanted by the sandwich terns and on several occasions I observed that a settlement of sandwich terns surrounded individual common terns and black-headed gulls so closely, that the latter deserted their clutch. These observations all refer to sandwich terns which settled at the edge of a black-headed gull or common tern colony. In the Griend situation the sandwich terns seldom settled in the middle of an already established black-headed gull colony, as described by SALOMONSEN (1947). He states (p. 184) that the sandwich tern on Hirsholmene (Denmark) '. . .forces its way to the centre of *L. ridibundus* colonies in the midst of the populated area, destroying the eggs and driving all birds away'. The description given by SALOMONSEN is in accordance with communications by the wardens of a number of colonies in the Netherlands, where sandwich terns also tend to settle in the middle of a black-headed gull colony. It thus appears that common terns and black-headed gulls usually have to make way for the sandwich tern. Space-competition, therefore, cannot be an important disadvantage of associative nesting as far as the sandwich tern is concerned.

Food-competition.

To a certain extent the food of the sandwich tern, the common tern and the black-headed gull overlap. Sandwich terns, however, are more specialised in their feeding habits than common terns and black-headed gulls and it should be stated that common terns and black-headed gulls feed on the same prey as do sandwich terns, whereas the reverse is only partly true. Sandwich terns mainly feed on a few species of pelagic shoal fish (I.2.4.1.). If a rich food source is detected, common terns and black-headed gulls immediately join the sandwich terns in fishing at the same place. This may well lower the sandwich tern's chances to catch a prey and, moreover, it implies black-headed gull predation on the fish caught. The sandwich terns in turn may react with joining the common tern if this pecies has detected a fish shoal. However, sandwich terns are restricted to a few food sources, whereas common terns and black-headed gulls may, and to a large extent do, exploit others. Therefore, in this aspect common terns and black-headed gulls will benefit more from the sandwich terns than the reverse.

4.3. THE ADAPTIVE VALUE OF DENSE COLONY NESTING IN THE SANDWICH TERN

Nest distribution pattern, anti-predator reactions and habitat selection of the sandwich tern are adapted to reduce predation on eggs and chicks. However, the question remains: why have sandwich terns evolved this type of nest-distribution; why do not they nest more spaced out, taking the benefit of an increase of camouflage of eggs and chicks which in combination with attacking a predator seems an effective defense of aerial and terrestrial predators both. The possession of such a defense mechanism would make the species less vulnerable to predation by ground mammals and less dependent on the presence of neighbouring common terns and black-headed gulls.

Many sea-bird species which nest on remote islands, nest at high densities. Thus, in an area which is free from mammalian predators, dense nesting seems of advantage to a great number of species. One cannot escape the impression that in the evolution of the nest distribution pattern of the sandwich tern, though it includes several anti-predator adaptations, factors other than predation must have played a major role. Among these factors are food and tidal movements.

Data on food and feeding habits have been given in I.2.4. Within the feeding area sandwich terns strongly react to conspecifics feeding successfully. Since nesting aggregations enable a species to hunt together in large numbers within a restricted area, it seems plausible that with regard to food colony nesting is of great advantage to the sandwich tern. But feeding habits cannot explain the high nest density which is so typical for the sandwich tern. However, synchronization of chick hatching might be of vital importance for the sandwich tern since it enables the species to profit maximally of transformation of information about food in the critical period of chick feeding. In the sandwich tern breeding is indeed highly synchronized. If this synchronization is promoted by a high density, as suggested for other dense nesting sea-bird species (NELSON, 1967, 1970; COULSON & WHITE, 1959), we might have an explanation for the high breeding density of the sandwich tern.

Besides feeding conditions, another factor may have promoted dense nesting as well. The clutch of the sandwich tern is in danger of being washed away by a high tide. To reduce the chance that eggs and chicks are washed away, the birds should nest on the highest places within the area. Usually, however, such places are scarce. Nesting at high densities enables the species to utilize the available space maximally. The following calculation for the Griend situation may serve as an example: On Griend 25.000 pairs of sandwich terns were recorded to nest in 1939. Within the same year, the area safe from high tides measured roughly 50.000 m^2. Taking into account that 80 to 90% of this area was probably unsuitable for nesting as a consequence of its vegetation and surface structure, roughly 5.000 to 10.000 m^2 ground must have been available for the terns. This would have allowed a nest density of 2.5 to 5 nests per square meter, which is in agreement with the actual density at which sandwich terns tend to nest. However, if the species should nest further apart than it actually does, in the above situation thousands of birds would have nested in constant danger of losing their eggs during a spring tide.

Summarizing the above suggestions it seems likely that the nest distribution pattern of the sandwich tern has evolved under the pressure of at least three different factors: food, predation and tidal movements. Dense colonial nesting may have promoted population synchronization which seems advantageous since it enables the species to profit maximally of transformation of information about food. Population synchronization causes a relatively large number of prey items to be available within a small period within a restricted area. If the predator population within this area is limited, this might be an important mechanism for reducing predation. In the nesting habitat of the sandwich tern the presence of mammalian ground predators is indeed limited; the area is flooded by high tides. However, this also implies that the clutch of the terns is in danger of being washed away. This in turn may have selected for special behaviour adaptations such as little or no attachment to the site of the previous year and selection of relatively high nesting places. Dense nesting has a negative influence on the camouflage of the brood. The sandwich tern seems to have abandoned several camouflage increasing behaviour aspects — the species defecates near the nest and does not remove empty egg-shells after the chick has hatched — in favour of avoiding exposure of the brood to predation. In combination with associative nesting this type of defense is very effective.

PART III

ASPECTS OF CAUSATION

1. INTRODUCTION

The temporal and spatial nest distribution pattern of the sandwich tern as well as the tendency to nest in association with other species have been shown to contribute to reproductive success. The present chapter deals with the behaviour mechanism through which the pattern of nest distribution is brought about. This question has also been taken up by PATTERSON (1965). In his study he showed that house-hunting black-headed gulls strongly prefer to settle in the most densely populated parts of the colony area. This behaviour causes clustering of nests. However, aggressive behaviour of territory owning individuals prevents house-hunting pairs from coming too close, which causes dispersion. The species-specific nest distribution pattern, he concluded, must be explained as a compromise between the opposing tendencies 'clustering' and 'spacing out'.

Although we had no arguments to expect that the spatial pattern of nest distribution in a sandwich tern colony is brought about in a way which is principally different from the situation in the black-headed gull, both species are markedly different with respect to territory size and to the way in which the territory is occupied. This made it interesting to study in the sandwich tern also how the spatial pattern of nest distribution comes about and what factors play a role in its determination. My approach has been to study the behaviour of the terns before and during settlement to explore the existence of a 'clustering' and a 'spacing out' mechanism. In order to investigate what factors influence the balance between both, special attention has been given to differences in the threat relationship of incubating, house-hunting and neutrally behaving individuals. The influence of size and density of a group on its attractiveness has been studied by manipulating the distribution of the terns on the island with the use of dummies.

The data which have been obtained with respect to the behaviour patterns which may have an influence on the timing of egg-laying are few and incomplete. Therefore, I have restricted myself to a brief summary of environmental influences on behaviour and behaviour interactions which can be expected to play a role in determining island and group synchronization in the sandwich tern.

The section on causal aspects of associative nesting with other species aims to find out whether associative nesting is the result of inter-specific attraction or simply the outcome of a similar habitat preference of different species. The possible ways in which associative nesting through inter-specific attraction might be brought about are discussed.

2. THE SPATIAL PATTERN OF NEST DISTRIBUTION

In order to investigate what behaviour patterns lead to the settlement of groups of sandwich terns and to the distribution of nests within such groups, the behaviour of the terns has been studied between arrival and egg-laying. Within this period on Griend the terns go through a number of successive and distinctly different stages (see also I.2.1.):

stage I : concentration on the tidal flats,
stage II : concentration in the nesting area,
stage III: settlement and egg-laying.

Unfortunately, in the pre-laying period sandwich terns are very sensitive to human disturbance. Several attempts were made to approach them within a distance that would enable detailed observations. However, these attempts all had the same result: the group disappeared and usually alighted somewhere else on the island. In order to get at least some insight in the activity of the terns in stage I, II and III, a number of behavioural units have been distinguished and the frequency of occurrence of these units have been determined for different groups, at different times of the day throughout the pre-laying period. To avoid disturbance the observations were all made from a distance of more than a hundred meters using a telescope (Hensoldt 40 × 60).

In stage III, after the first eggs have been laid within the group, the terns gradually become less sensitive to disturbance. In this situation in 1970 two groups could be approached with a hide up to a distance of a few tens of meters without causing noticeably disturbance. In both groups the majority of pairs had already begun laying, but there were still several pairs without eggs. In these groups several house-hunting pairs have been watched over a period of several days, until the first egg was laid. Special attention has been given to aggressive encounters between house hunters and other birds present in the group.

2.1. ACTIVITY CHANGES THROUGHOUT THE PRE-LAYING RERIOD

Description of the behavioural units.

A number of behavioural units have been distinguished. Most units consist of a single posture, which may or may not include a stereotyped movement and/or calls.

Certain units, however, (*e.g.* circling, copulation) are composed of a complex of movements and postures which together are the components of a fixed pattern.

(1) *Stretch posture* (Fig. 36a)—The neck is vertically stretched and the carpal joints are raised. During a 'high intensity stretch' the body axis points upwards at an angle up to 45° to the horizontal; the tail may be raised at the same angle to the horizontal. In general the head is held horizontally, but in certain situations—*e.g.* when a bird offers a fish in a stretch—head and bill are bent downwards (fig. 36b).

(2) *Circling* (Fig. 36c)—Two birds walk around each other in the stretch posture. The body axes roughly parallel each other and the birds are in a head to tail position with respect to one another. The carpal joints are conspicuously raised and sometimes hook each other. As a consequence both birds may walk 'arm in arm'. The whole procedure may last from a few seconds up to several minutes.

(3) *Erect posture* (Fig. 36d)—The erect posture includes all characteristics of a 'high intensity stretch posture' with the only exception that the body axis points more upwards (up to 60° with the horizontal) while the head is in line with the neck.

(4) *Stooping* (Fig. 36e)—The body axis points downwards up to 40° with the horizontal. The neck is held at an angle of roughly 90° with the body axis. The head is turned downwards and the bill usually points to the ground. In general the carpal joints are raised.

(5) *Fixating* (Fig. 36f)—The body axis is horizontal. The neck is held in line with the body axis, head and bill point vertically downwards. The bird stands motionless fixing the bottom.

(6) *Scraping* (Fig. 36g)—The bird lies with its breast on the ground. With a backward scraping movement of the legs sand is thrown away from under the belly.

(7) *Advertisement calling* (Fig. 36h)—The body axis points slightly upwards (up to 30° with the horizontal). The neck is withdrawn, carpal joints, crest and body feathers are conspicuously raised. Head and bill point into the direction of a fish-carrying conspecific flying around. The bird calls loudly and maintains its position with respect to the fish-carrying individual by turning around.

(8) *Begging* (Fig. 36i)—The bird rhythmically calls in a hunched position. The body axis is horizontal, head and neck are withdrawn between the shoulders. At every call head and bill rhythmically move up and down between 0° and 30° with the horizontal.

(9) *Threatening* (Fig. 36j)—Head and neck are stretched into the direction of another bird. The carpal joints, crest, neck and body feathers are raised. In 'high intensity threat' head and neck rhythmically move up and down. Threatening occurs in sitting as well as in standing position.

(10) *Fighting*—In threat position both opponents seize each others bill and thereupon bite, twist and pull. When fighting in standing position, the wings are often raised. The wings, however, are never used to beat the opponent.

(11) *Copulation* (Fig. 36k)—The body of the female is between horizontal and sloping slightly downwards. The neck is withdrawn; head and neck point upwards at an angle between 30° and 80° with the horizontal. During copulation the female may peck at the breast of the male. The body axis of the male makes an angle of roughly 45° with the horizontal. The legs are bent and the wings spread. The neck is stretched in an upward direction; head and bill point downward. During cloacal contact the male beats his wings.

In addition to the above described units, standing, walking, sitting and preening have been distinguished.

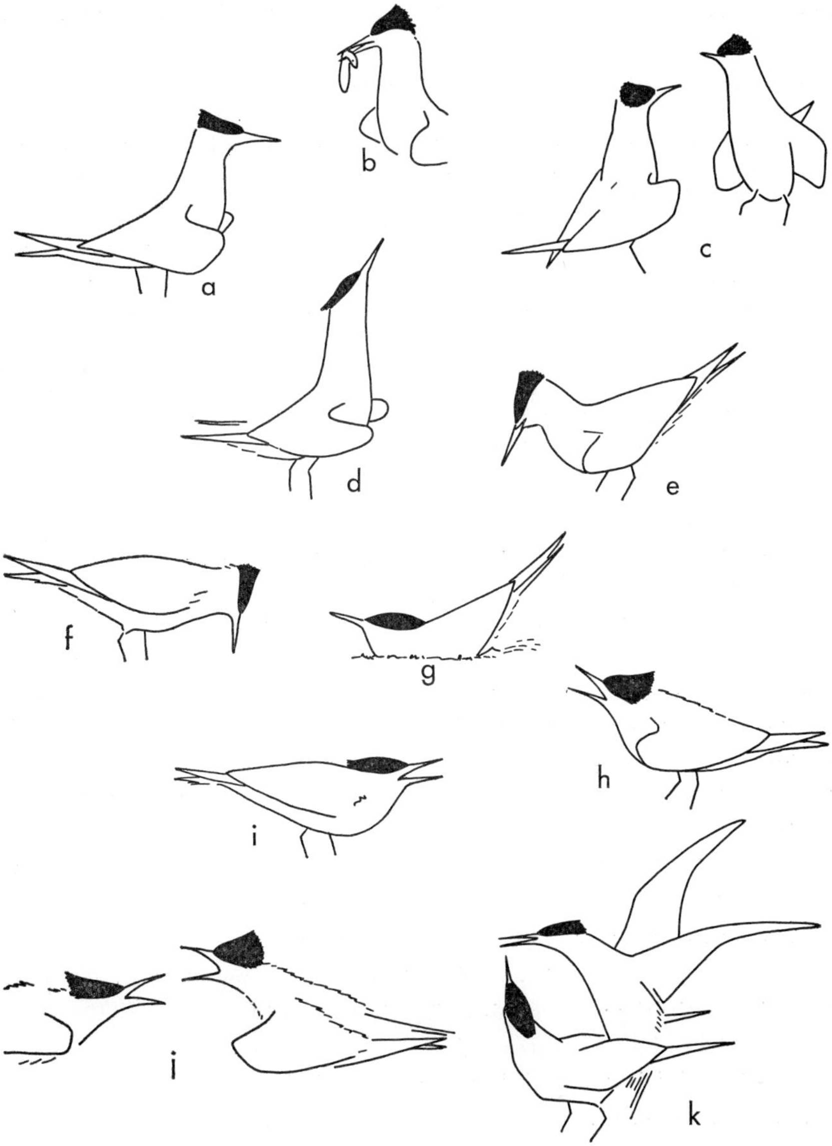

Fig. 36. a,b. Stretch posture; c. Circling; d. Erect posture; e. Stooping; f. Fixating; g. Scraping; h. Advertisement calling; i. Begging; j. Threatening; k. Copulation.

The context in which the various units are shown.

The majority of the terns of the Griend population are probably mated when they arrive in the breeding area. However, in the beginning of the season there also were several unmated individuals. Among the latter the females tend to concentrate in groups on the tidal flats. These groups are visited by fish-carrying males giving a special call ('fish-call', VAN DEN ASSEM, 1954c). To this call the females may react with advertisement calling. When a fish-carrying male alights near a female, both exhibit the stretch posture; the fish may be transferred and thereupon the erect posture and circling may be shown. After a while both birds usually fly up and either disappear to the feeding area or start displaying in the air. (A short description of aerial displays has been given by CULLEN, 1960b.)

In the beginning of the season aerial displays are prominent among both mated and unmated individuals. However, when pair-formation progresses ground displays gradually take their place. During greeting ceremonies the fish supplied by the male is transferred more and more rapidly, stretch and erect postures take less time, circling becomes more prominent and the whole procedure increasingly leads via begging to copulation.

Still later the first signs of an awakening interest in the surface features of the nesting area appear. The birds start walking in pairs over the sandy ridge of the island, inspecting suitable nesting places. For the first time stooping and fixating are shown, later on interspersed with scraping.

Threatening and fighting as well as walking, standing, sitting and preening may be shown throughout the season in nearly all situations.

Based on the functional context in which the various behaviour units are shown, they have been grouped as follows:

(1) courtship activities: stretch posture, circling, erect posture, advertisement calling, begging, copulation,
(2) nesting activities: stooping, fixating, scraping,
(3) overt aggression: threatening, fighting,
(4) neutral activities: walking, standing, sitting, preening.

Activity in relation to stage of settlement and group density.

Table 51 gives for groups of terns in the various stages distinguished the proportion of birds involved in different activities. Sometimes birds in the end of stage II and in the beginning of stage III occurred so close

TABLE 51

Proportion of birds involved in different kinds of activities in relation to stage of settlement. Further explanation in text

Stage Place of concentration	I Tidal flats	II Dunes 'South'	II + III Dunes 'South'	II + III Dunes 'West'	III Dunes 'North'
Courtship activities	5.8%	12.0%	14.7%	18.0%	4.1%
Nesting activities	0.0%	0.5%	3.3%	4.0%	3.4%
Overt agression	0.0%	0.9%	3.5%	4.2%	13.6%
Neutral activities	94.2%	86.6%	78.5%	73.8%	78.9%
Number of birds on which percentages are based	629	4797	3724	765	3756
Observation period	18-27 April		1-8 May		19-22 May

together that their distinction during the observations was no longer feasible; consequently they have been taken together. The table shows a peak of courtship activities at the end of stage II. Nesting activities are relatively high at the end of stage II and in stage III. Overt aggression reaches a peak in stage III.

Besides differences in activity, the various groups of terns observed were also different with respect to group density. Unfortunately, the observation distance did not allow a good estimate of the group density. On average, in stage I, II and III individual birds or pairs were spaced apart more than 100 cm, 50-100 cm and 40-50 cm, respectively.

Summarizing we conclude that the terns of the Griend population go through three successive stages which are characterized by an increase of group density and an increase of the proportion of birds occupied in nesting activities and overt aggression. The proportion of birds involved in courtship activities reaches a peak at the end of stage II, *i.e.* the stage prior to settlement and egg-laying.

It is tempting to explain the increase of aggression, courting and nesting as the result of the increased tendency to cluster mediated by what is known as social facilitation. However, a causal relationship in the reversed sense is possible: an increase of nesting causing through a common interest in suitable nesting places an increase of the group density, and an increase in aggression and courting reducing the distance between individual birds. Although this explanation seems much less likely to me, it has never been disproved.

2.2. FACTORS DETERMINING INTER-NEST DISTANCES

As mentioned before, to minimize disturbance, short distance observations on sandwich terns occupied in nest-site selection were made several days after a particular group had settled in the area. In all cases the observations were started when the majority of pairs had already begun laying. As a consequence the house-hunting pairs under observation all belonged to the 'tail' breeders of the group.

During the observations three categories of birds have been distinguished:

(1) incubating birds,
(2) house hunters — birds involved in activities earlier described as 'nesting activities',
(3) neutral birds — birds involved in activities earlier described as 'neutral activities'.

The house-hunters were usually present in pairs. The neutral birds were either single or in pairs. The latter category probably included several birds which should be regarded as potential house-hunters.

Clustering.

During the observations it soon became clear that the house-hunters nearly always stayed in the vicinity of nest-owning conspecifics. They often made prolonged inspection visits along the colony border, showed special interest for the surface features in the neighbourhood of nesting conspecifics and often attempted to intrude between nests situated in less dense parts of the group. At different places a nest scrape might be made. In the course of time they gradually lost their interest in most of these scrapes and finally concentrated on only one, in which the eggs was laid. Figure 37 gives for a number of house-hunting pairs present in colony N II (1970) the area in which these birds concentrated their nesting activities, the places where they attempted to intrude between nests and the location of scrapes. The figure clearly demonstrates the strong tendency of the house-hunters to select a nesting site within a very short distance of incubating conspecifics.

Spacing out.

The presence of house-hunting sandwich terns at short distances of incubating conspecifics nearly always evoked threat in the latter. In order to collect information on the size of the area defended by the incubating

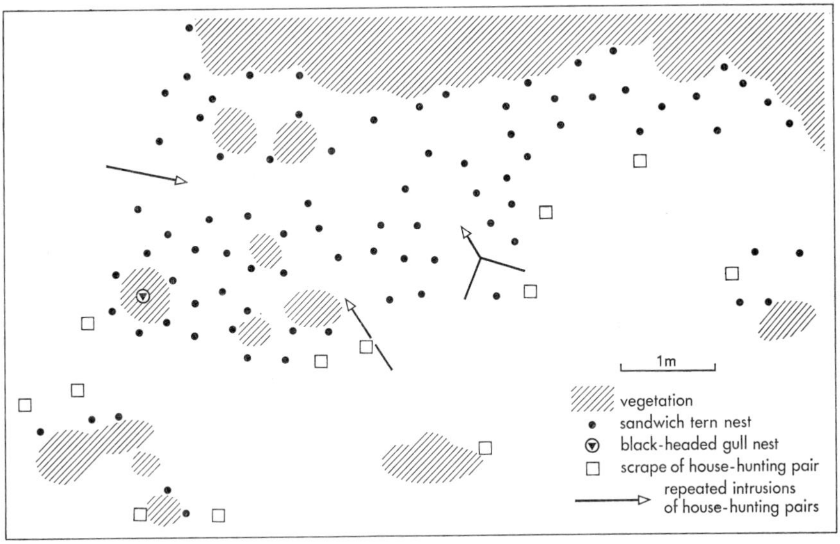

Fig. 37. Part of the study area (colony N II) where house-hunting pairs concentrated their nesting activities.

birds, around a number of nests circles were painted with a radius of 50 cm. This made it possible to estimate the distance at which sandwich terns walking in the direction of a nest were threatened by the incubating birds.

Figure 38 (curve a) gives for different distances from the nest the percentage of cases in which a non-incubating intruder was threatened by an incubating bird. It shows that the radius of the area defended by the incubating birds averaged 57 cm. The figure also gives the distance at which neutrally behaving intruders in reaction to threat stopped approaching (curve b). These birds clearly reacted to a threatening incubating bird with avoidance behaviour. Finally, Figure 38 gives the frequency distribution of inter-nest distances of the 'tail' (curve c) and 'peak' breeders (curve d) of a number of groups observed in 1970. A comparison of curve c and d with a and b shows that both peak and tail breeders selected a nesting site which, in general, lays well within threat distance of the nearest incubating neighbour. They also nested within the distance at which the neutrally behaving birds reacted to threat with avoidance behaviour, which means that the house-hunters, in contrast to the neutral birds, persisted in approaching the incubating birds despite the threat of the latter.

In order to be able to select a nesting site at short distance of an

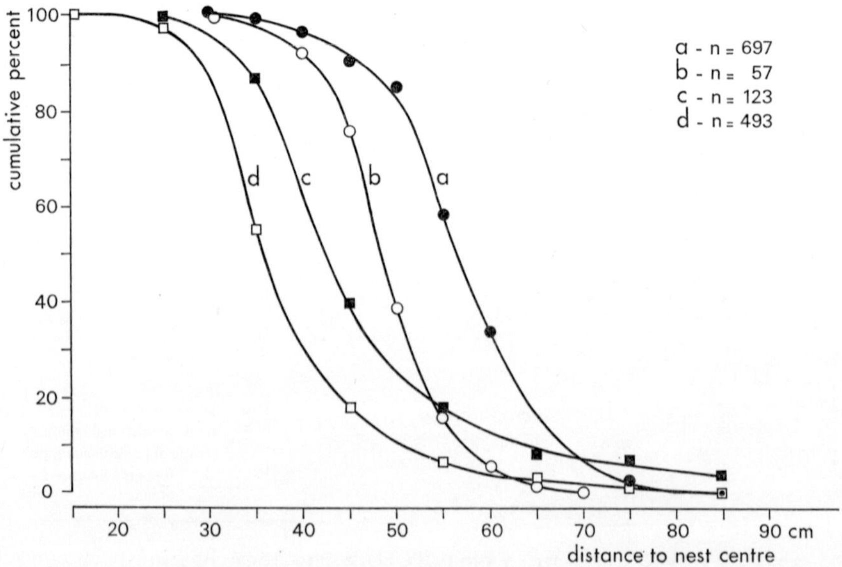

Fig. 38. Frequency distribution of (a) distances at which incubating birds threatened and approaching intruder, (b) distances at which neutrally behaving intruders in reaction to threat stopped approaching, (c) minimum inter-nest distances of 'tail' breeders, and (d) minimum internest distances of 'peak' breeders. Further explanation in text.

already incubating bird, house-hunting pairs can be expected to prefer the neighbourhood of relatively less aggressive incubating birds for nest site selection. To test this, we first have to establish if aggressiveness of incubating birds measured as 'the radius of the area defended against intruders' is independent of the number of intrusions made by non-incubating birds. Figure 39 gives the radius of the area defended by the incubating birds in relation to the number of intrusions observed during the observation period. The frequency of intrusion seems not to have influenced the average distance at which an incubating bird threatened an approaching conspecific. This means that the resistence of an incubating bird does not importantly change under the influence of the persisting intrusions of a house-hunting pair. To investigate the possible preference of house-hunting pairs to concentrate their nesting activities near less aggressive incubating birds, those incubating birds near which house-hunting pairs repeatedly visited a scrape have been indicated in Fig. 39 by an open dot. No preference of house-hunting birds for incubating birds of particular aggressiveness is apparent from the figure.

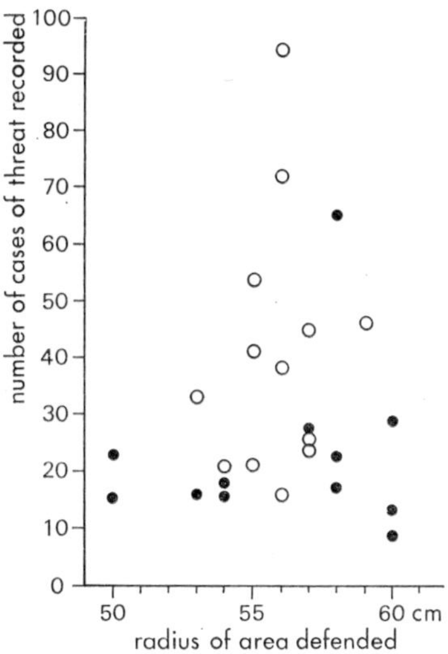

Fig. 39. The number of cases of threat observed, as an indication of the incidence of intrusions by house-hunters, compared with the size of the area defended.

The likelihood of responding aggressively to nearby conspecifics varied considerably between incubating, house-hunting and neutral birds. Data on the occurrence of threat collected during aggressive encounters are summarized in Figure 40. It shows that house-hunters and neutral birds were often threatened by incubating birds. However, the house-hunters and the neutral birds rarely threatened an incubating bird, but they often threatened one another.

Differences in intensity and duration of threat followed the same trend as differences in threat frequency. Besides, in all cases in which house-hunting and neutral birds threatened an incubating bird this was in reaction to threat started by the latter.

Marked differences were observed in the behaviour of house-hunters and neutral birds in reaction to a threatening incubating bird. The house-hunters usually continued their nesting and courtship activities within short distances of the threatening birds and only reacted by keeping out of reach of the bill of the threatening bird. The neutral birds showed much stronger avoidance behaviour; they immediately walked away from the threatening birds.

In addition to the aggressive encounters represented in Figure 40, we often witnessed furious and long-lasting aggressive encounters between neighbouring house-hunting pairs, between neutral birds and

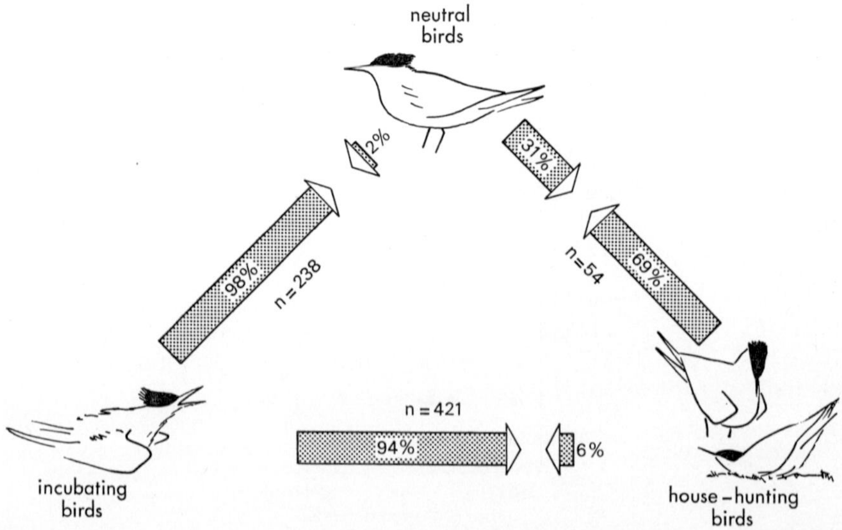

Fig. 40. Occurrence of threat (%) during aggressive encounters between incubating, house-hunting and neutrally behaving birds. The arrows indicate the direction of threat.

between incubating birds. So, all categories of non-incubating birds may be threatened frequently by both incubating and non-incubating birds, but the incubating birds are rarely threatened by non-incubating birds. This suggests that something in the behaviour of the incubating birds inhibits threat of non-incubating individuals. In this respect it is of interest, that the majority of cases in which house-hunters threatened an incubating bird coincided with standing up in the latter. We also found a relatively high threat frequency of house-hunters towards incubating birds, when the mate of the incubating bird was standing at the nest rim (these cases have not been included in the data presented in Fig. 40). It thus seems possible that the sitting position of an incubating bird inhibits threat of non-incubating individuals.

Differences between peak and late breeders.

It has been shown before (II.1.3.3.2. and Fig. 38) that the peak breeders of a group nest much closer together than the tail breeders. In the light of what we know about the factors determining inter-nest distances, such differences can only be explained by differences in the relative strength of the opposing tendencies 'clustering' and 'spacing out' at the moment of nest site selection. Nest site selection of peak and late breeders takes place under different circumstances: Among the peak breeders, neighbouring pairs usually are in the same phase of breeding. This means that house hunting pairs have to compete for a site with other house-hunting pairs; a process of long lasting mutual aggressive interactions. In contrast, the tail breeders of a group usually settle in the neighbourhood of an already incubating bird to which they behave in a markedly submissive way.

To investigate how differences in phase of breeding between neighbouring pairs are related to inter-nest distance, for a number of nests the distance to the nearest neighbour at the moment of egg-laying has been plotted against the difference in laying date in days. Figure 41 gives the results. It shows that laying date differences of 0-3 and 4-20 days coincide with a median inter-nest distance of 35,5 and 47,5 cm, respectively. Within the range of 0-3 days and within the range of 4-20 days median inter-nest distances were not different.

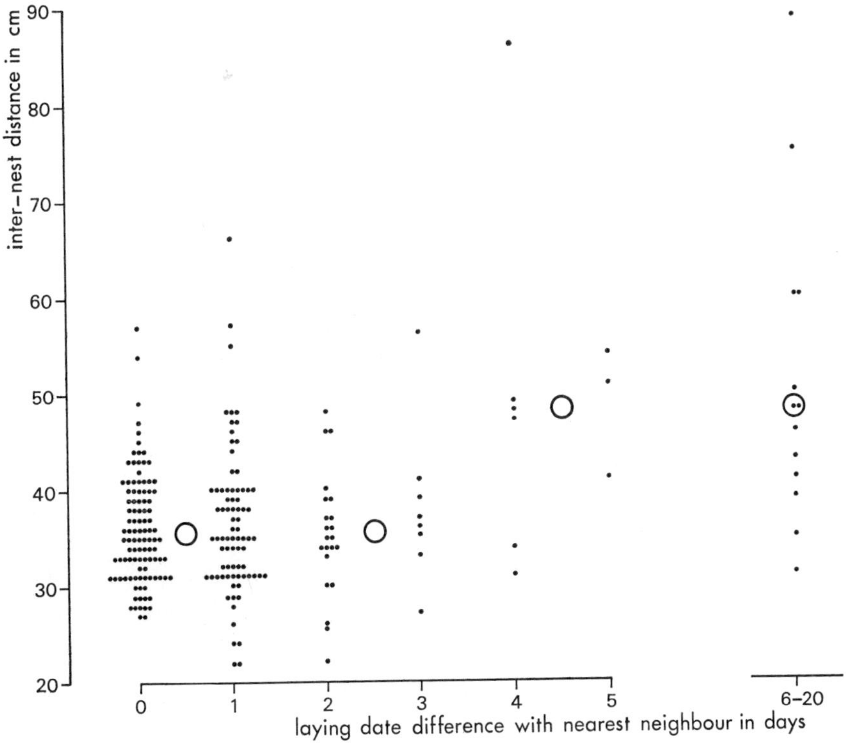

Fig. 41. Relationship between differences in laying date of neighbouring sandwich terns and the distance between their nests. The open circles indicate the mean of class 0 + 1, 2 + 3, 4 + 5 and 6-20, respectively.

The relatively high but consistent value of median inter-nest distances of pairs with a breeding phase difference of more than 3 days is a point of special interest. The period of nest-site selection measures 3 days on average. A breeding phase difference of less than 3 days thus means that neighbouring pairs are in the house-hunting phase at the same time for at least part of the period of nest-site selection. Breeding phase differences of 4 days or more implicate that nest site selection takes place in full presence of a neighbour which is incubating.

Summarizing we conclude that in the sandwich tern two factors are responsible for the inter-nest distances within a group: house hunting pairs select a nesting site at very short distance of nesting conspecifics (clustering), but territorial aggression prevents them from coming too close (spacing out).

There is a special relationship between incubating and non-incubating individuals: non-incubating individuals are very often threatened but non-incubating individuals seldom threaten incubating individuals. When selecting a nest site house-hunting pairs behave in a markedly submissive way towards incubating individuals. The sitting position of incubating individuals seems to inhibit aggression of house-hunters.

Differences in minimum inter-nest distance between peak and tail breeders of a group might be explained by differences in the phase of breeding of neighbouring birds at the moment of nest-site selection.

2.3. EXPERIMENTS WITH SANDWICH TERN DUMMIES

2.3.1. The influence of size and density on group attractiveness.

Observations in a group of sandwich terns a few days after settlement have taught us some factors determining inter-nest distances. However, we still know very little about the factors which determine the attractiveness of a group for house-hunting pairs. In order to obtain information about the possible significance of size and density of a group for its attractiveness, between 1967 and 1971 a number of experiments have been done with dummies.

Two kinds of dummies were used: dummies representing a sandwich tern and dummies representing a black-headed gull, all in sitting position, of natural size and painted in natural colours (Plates 23 and 25). In 1967 the dummies were made of gypsum. In 1968, 1970 and 1971, in addition to the gypsum ones, dummies of polyester were used. At some distance gypsum and polyesther dummies could only be distinguished by a slight difference in colour.

In all years of observation the dummies were placed in squares of 15 × 15 m inconspicuously marked with pieces of drift wood. The distance between neighbouring squares was also 15 m. In 1967, 1968 and 1970 the dummies remained in the squares from the beginning until the end of the experiment. In 1971 they were removed after each test. Observations were made from a hide on a platform two meters above the ground at a distance of more than a hundred meters from the area of the experiment. Sandwich terns present in the neighbourhood of the dummies were continuously followed. Within the squares the number of terns was counted every five minutes. To avoid the possible influence of breeding birds nesting in the neighbourhood of the dummies, the experiments were carried out in a rather bare and homogeneous area unsuitable as a nesting place for sandwich tern, black-headed gull or common tern. The possible influence of terrain differences between the squares was avoided by alternately placing the dummies in each square. Observations were done in April and May, mainly between 18.00 and 20.00 hours.

In the following account the presence of sandwich terns in the area of the experiment has been expressed in four different ways:

(1) Intention landings: Intention landings represent those cases in which a sandwich tern fluttered above one of the squares in a way typical of a bird about to land. They were only scored if there were no other terns present in the experimental area.

(2) First landings: A first landing was scored if a sandwich tern alighted in one of the squares, while at the same time no other terns were present in the area. The first landings thus are indicative for the choice of the terns.

(3) Average number of birds present: For each square the average number of terns present has been calculated over the total number of counts in which a positive score was made in that square.

(4) Rate of occupancy: The proportion of the total number of counts in which at least one sandwich tern was present within a certain square has been taken as a measure for the occupation of that square.

Experiment I.

In 1967 20 sandwich tern dummies and 20 black-headed gull dummies were simultaneously presented to test the usefulness of both dummies. The dummies were placed to imitate a homogeneous (one-species) colony. In one square (Figure 42a) only sandwich tern dummies were placed at distances of 50 cm; in a second one black-headed gull dummies spaced apart 1-2 m, while a third square remained empty. The black-headed gull dummies were also given an artificial nest of grass and twigs. Observations were made between 6 and 15 May.

During the observations it soon became clear that sandwich terns flying over the area showed great interest in the dummies of their own species. They often alighted in the neighbourhood of these dummies

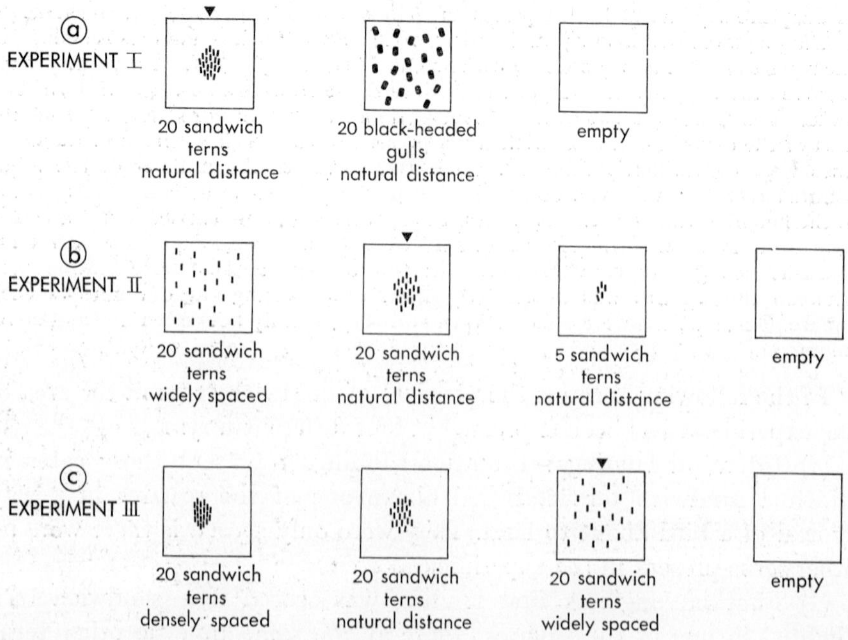

Fig. 42. The position of sandwich tern and black-headed gull dummies during a number of experiments in which the influence of group size and density on group attractiveness was tested. The arrows indicate the groups of dummies preferred by the terns.

(Plate 20 and 21) and several times they attempted to copulate with their gypsum 'conspecifics'. Sometimes the number of terns present in the area strongly increased in the course of a test. In such a situation many of them might walk around, also visiting the squares next to the one in which the sandwich tern dummies had been placed. Neither common terns nor black-headed gulls were recorded in the area during the experiment. The results are summarized in Table 52a. The relatively high number of first landings in the squares in which the tern dummies had been placed clearly indicates that the terns were attracted by the dummies of their own species. The average number of birds present and rate of occupancy were also highest in the squares in which the tern dummies had been placed. However, these measures are probably not independent of the first landings, for it is likely that the presence of one living tern influences the presence of others. Of course, we cannot conclude from these results that sandwich terns do not react to black-headed gull dummies. The reactions of the terns to the sandwich tern dummies might have hidden their interest in the gull dummies.

Plate 19. Sandwich terns in the stretch posture.

20

2

TABLE 52

Reactions of sandwich terns to sandwich tern (ST) and black-headed gull (BHG) dummies placed in groups varying in size and density

(a)

	20 ST (natural distance)	20 BHG (natural distance)	Empty square
First landings	33	2	4
Average number of terns present	4.5	0.8	1.4
Rate of occupancy	45.6	5.3	5.4

(b)

	20 ST (widely spaced)	20 ST (natural distance)	5 ST (natural distance)	Empty square
First landings	3	14	2	0
Average number of terns present	2.0	13.7	2.9	0.0
Rate of occupancy	0.7	8.0	3.8	0.0

(c)

	20 ST (densely spaced)	20 ST (natural distance)	20 ST (widely spaced)	Empty square
First landings	1	3	7	1
Average number of terns present	1.0	1.2	2.0	2.0
Rate of occupancy	0.2	1.0	2.1	0.2

(d)

First landings	11 (21)	16 (28)	2 (5)	0 (0)
Average number of terns present	29.5	6.1	2.0	0.0
Rate of occupancy	13.9	8.2	2.1	0.0

Experiment II.

In 1968 four different squares were pegged out in which the following groups of tern dummies were placed (see Plate 22 and 23): One group of 20 dummies at 1-2 meters from one another, one group of 20 dummies at 50 cm from one another and one group of 5 dummies at 50 cm from one another (Figure 42b). Observations were made between April 26 and May 5. The results are given in Table 52b. The table shows that the num-

Plates 20 and 21. In 1967 during an experiment with sandwich tern and black-headed gull dummies, the terns only alighted near the dummies of their own species. Both plates show terns and dummies; the sitting birds are all dummies. (phot. J. van de Kam).

ber of first landings was much higher in the squares in which the 20 'normally spaced' dummies had been placed than in any of the other squares. If we compare the number of first landings near the 20 normally spaced dummies with the number recorded near both other groups, our results indicate that a larger number of dummies and a normal spacing is preferred over a smaller number and a widely spaced group. Thus, number and density of a group both determine its attractiveness.

Experiment III.

In order to analyse the influence of group density further, another four squares were pegged out in 1970. Three groups of 20 sandwich tern dummies each, spaced apart 25 cm, 50 cm and 1-2 meter were simultaneously presented (Figure 42c). Observations were made between April 13 and May 11. The results are given in Table 52c. It follows that the majority of first landings was recorded near the '20 wide' group of dummies, which is contradictory to what we had expected. However, shortly after the start of the experiment a large group of sandwich terns concentrated at a distance of a hundred meters from the squares. The first landings recorded all took place in presence of this group of terns and, except for one, they were all recorded from the square which was nearest to these terns. It is likely, therefore, that the living terns strongly influenced the results obtained.

Consequently, in 1971 the experiment was repeated. Observations were made between April 29 and May 11. The results are given in Table 52d. The numbers of first landings recorded near the groups '20 dense' and '20 normal' were much higher than those near the '20 wide' group or in the empty squares. These results once more indicate that a relatively dense group of dummies is preferred over a widely spaced group. The difference between the '20 dense' and '20 normal' groups of dummies is rather small. However, the number of intention landings recorded near both groups show a picture corresponding to that of the first landings, which supports the conclusion that the terns prefer to alight near the '20 normal' group of dummies. The 'average number of birds present' and the 'rate of occupancy' of the squares in which the '20 dense' group of dummies had been placed are — in comparison with the other squares — relatively high. However, these higher numbers were due to one large group of terns which stayed near the '20 dense' group of dummies for a relatively long time. Consequently, these results do not disprove the above indication that the sandwich terns preferred to alight near the '20 normal' group of dummies.

In order to have a better idea of the way in which the terns responded to the dummies it seems worthwhile to discuss the reactions of the terns in some detail.

In all years of observation the number of terns that appeared in the area of the experiment was only a very small proportion of the total number present on the island. Besides, the number of terns visiting the dummies strongly decreased in the course of the season, which might be explained by (a) habituation and (b) an increase of the number of terns which had already settled on the island.

There were marked differences in the reactions of the terns in 1967 and in the years following. In 1967 the terns strongly responded to their gypsum 'conspecifics' as demonstrated by the extremely high rate of occupancy in that year (Table 52a). In 1968, 1970 and 1971, however, only a few terns were recorded in the area of the experiment and on several days no landings at all were recorded. In 1967 we also noticed several copulation attempts with the dummies, whereas this was never recorded in 1968, 1970 and 1971. The 1967 experiment differed from the experiments in 1968, 1970 and 1971 in the following aspects:

(a) In 1967 the dummies were all made of gypsum, whereas in the other years most of them were made of polyester. Although we never got the idea that once the terns had alighted, they reacted differently to both types of dummies, differences between the polyester and gypsum dummies — for instance in the roughness of the surface — may have influenced their attractiveness for the terns.

(b) To avoid competition with living terns, the dummies were always presented in an area unsuitable for nesting. Since the distribution of sandwich terns on the island varied from year to year, different places were chosen for the experiments. Differences in the physical character of the areas chosen in 1967 and in 1968, 1970 and 1971 were small. As a consequence, an influence of such differences, though it cannot be excluded, is not very likely.

(c) In 1967 black-headed gull dummies were used in addition to the tern dummies. The presence of black-headed gull dummies in 1967 may have influenced the attractiveness of the nearby tern dummies.

2.3.2. The nature of the reactions to the dummies.

Was alighting near conspecifics a social response in general or a response in particular motivated by a tendency to incubate? We felt that the latter could be taken to hold if the settlement of the terns on the island could be manipulated by the placement of dummies. In 1968, 1970, 1971 and 1972, after the above described experiments had been finished, groups of dummies were placed in an area which to my opinion was a suitable but not an optimal nesting place for sandwich terns. These groups consisted of 5 or 20 dummies arranged in a way imitating as much as possible a group of nesting sandwich terns. At 3-day intervals a check was made to establish whether or not sandwich terns had settled in the neighbourhood of the dummies. Each nest was marked with a wooden peg and distances between nests and dummies were measured after the breeding season.

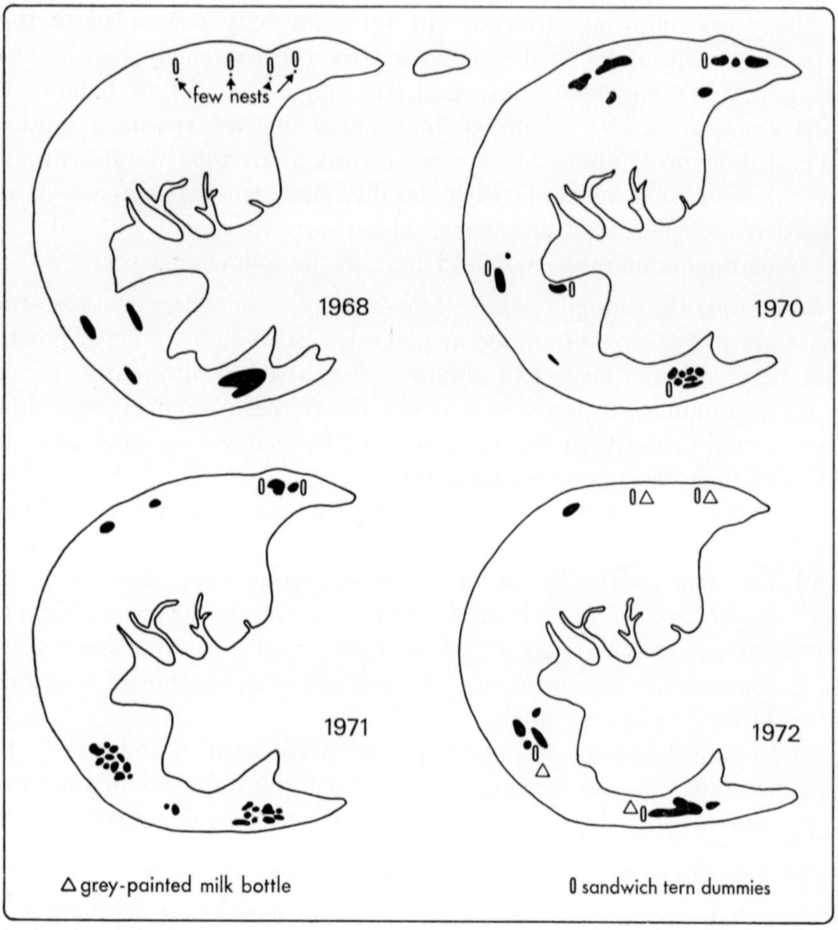

Fig. 43. Position of groups of sandwich tern dummies and grey-painted milk bottles together with the distribution of sandwich tern colonies on Griend in 1968, 1970, 1971 and 1972.

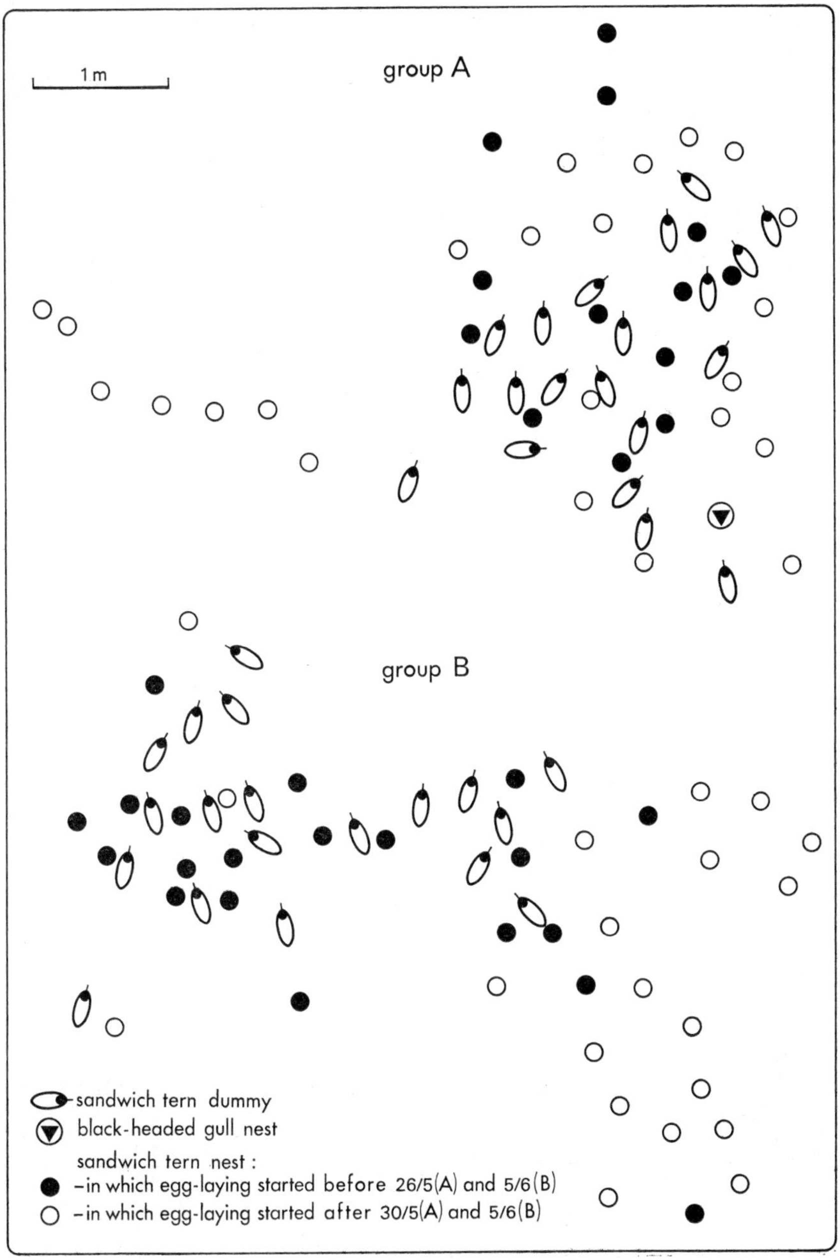

Fig. 44. Position and date of egg-laying of sandwich tern nests situated in the neighbourhood of the dummies.

Figure 43 gives the distribution of groups of dummies and sandwich tern colonies on the island. In all years of observation sandwich terns settled in the neighbourhood of the dummies (Plate 24). The number of nests that appeared near the dummies was small in 1968, but large numbers were recorded in 1970, 1971 and 1972.

Figure 44 gives the spatial distribution of nests and the date of egg-laying of two groups of terns which settled near the dummies in 1970. The figure shows that some terns nested at short distance from the dummies. In both groups the terns which had settled nearest to the dummies started laying relatively early.

From the above results we conclude that in the sandwich tern the selection of a nesting site can be manipulated by placing gypsum and polyester dummies in the breeding area. Alighting near the dummies thus seems a response which is motivated by a tendency to incubate. The influence of the dummies on nest site selection was especially clear among the first nesting birds of a group. Those which settle later may be attracted by the dummies as well as by the living birds already present.

The occurrence of copulation attempts during the first series of experiments, together with the fact that the terns nested near the dummies provide strong evidence that the dummies 'to some extent' were reacted to as nesting conspecifics. However, it seemed worthwhile to test once more the specificity of the dummies by placing them in the area together with other objects of the same basic colour, size and group density, but lacking the species' specific shape and colour pattern. For this purpose on 5 May 1972 sandwich tern dummies and milk bottles painted grey were placed on the island. At four different locations a group of five dummies was placed at a distance of 5-10 meter from a group of 5 bottles. Within a group dummies and bottles were spaced apart 50 cm. In two cases sandwich terns settled in the area of the experiment (see Fig. 43) Figure 45. gives the spatial nest distribution pattern for both situations. Eggs were laid at short distances of the dummies; no nests were found near the bottles. Although the number of tests with a positive reaction are few, these results provide another indication that during the process of nest selection sandwich tern dummies are reacted to as nesting conspecifics.

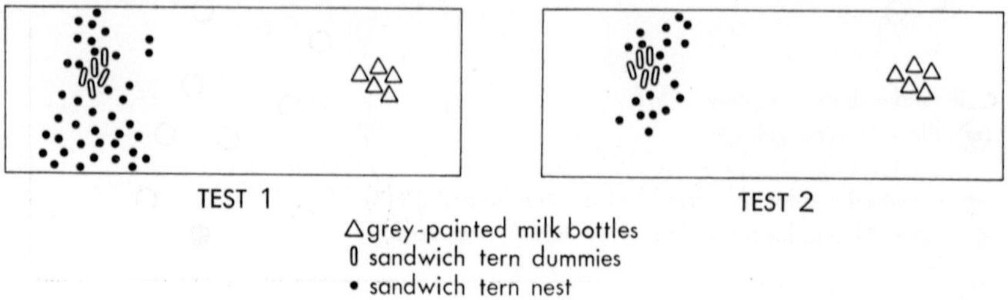

Fig. 45. Position of sandwich tern dummies, grey-painted milk bottles and sandwich tern nests during the 'dummy-bottle' experiment. For explanation see text.

Since in the natural situation the selection of a nesting site is a complicated procedure which through many social interactions leads to a balanced pattern of inter-nest distances, it seemed interesting to study the nest position of the terns which had settled near the dummies.

Figure 46 gives the frequency distribution of minimum distances between dummies and nests together with the distribution of minimum inter-nest distances in 1970 and 1971. The figure shows clearly that on average the dummy-nest distances were smaller than the inter-nest distances. (Plate 24).

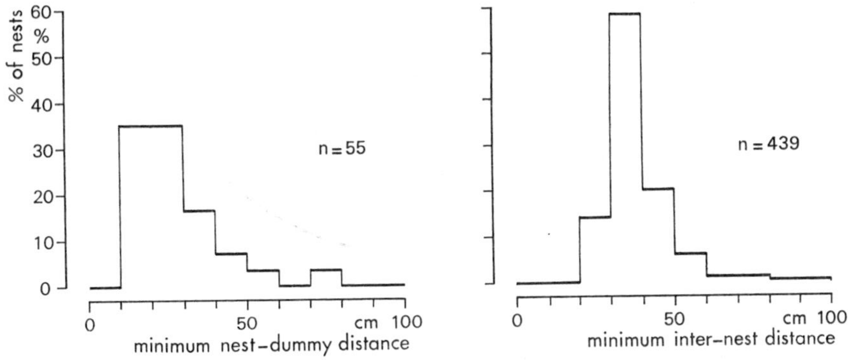

Fig. 46. Frequency-distribution of minimum distances between sandwich tern dummies and sandwich tern nests (left) and between neighbouring sandwich tern nests (right) in 1970 and 1971.

Figure 47 gives the position of sandwich tern nests in relation to the dummies. Only a small proportion nested in front of the dummy's head.

The relatively short distances between nests and dummies might be explained as a consequence of the presence of a 'clustering mechanism' in the absence of a 'spacing-out mechanism'. The preference of the terns to settle sideways of the dummies might be explained by avoiding the potentially dangerous bill.

3. TIMING OF EGG-LAYING

This section deals with the question of what factors are responsible for the timing of egg-laying which in the sandwich tern leads to island and group synchronization.

3.1. ISLAND SYNCHRONIZATION

In the sandwich tern egg-laying takes place within a three month period and within this period by far the majority of eggs is laid within

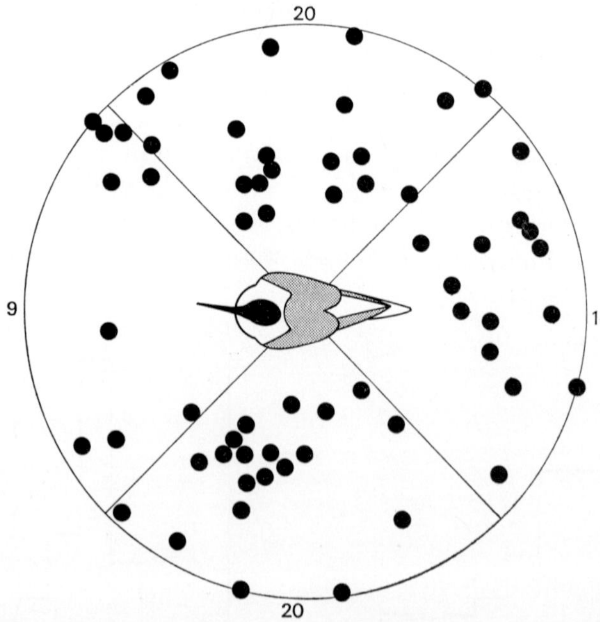

Fig. 47. Position of sandwich tern nests in relation to the dummies. The black dots indicate the centre of a nest. For explanation see text.

only a few weeks. Figure 48 gives in outline the distribution of egg-laying of the Griend population. (The figure has been based on data collected in 1971 and 1972, see also Fig. 8.) Two different peaks of laying can be distinguished. Both peaks are markedly different with respect to seasonal timing of laying (a-a′) as well as to degree of synchrony (b-b′)

In several bird species the onset of migration has been shown to be influenced by abiotic environmental factors, among which daylength seems of special importance. The influence of these factors upon the individual seems dependent of a fixed internal cycle as well as on gonadal development. Ring recoveries (MÜLLER, 1959) have shown that sandwich terns do not return from their African winter qaurters to their European breeding grounds until they are sexually mature at the age of 3-4 years. It has also been shown that recruit breeders arrive later in the area and during a longer period than recidivist breeders. Thus, in the sandwich tern differences in time and spread of arrival between recidivist and recruit breeders can be expected to lead to differences in laying date and degree of synchrony between both groups. Unfortunately, our data on the arrival of the recruit breeders are insufficiently complete

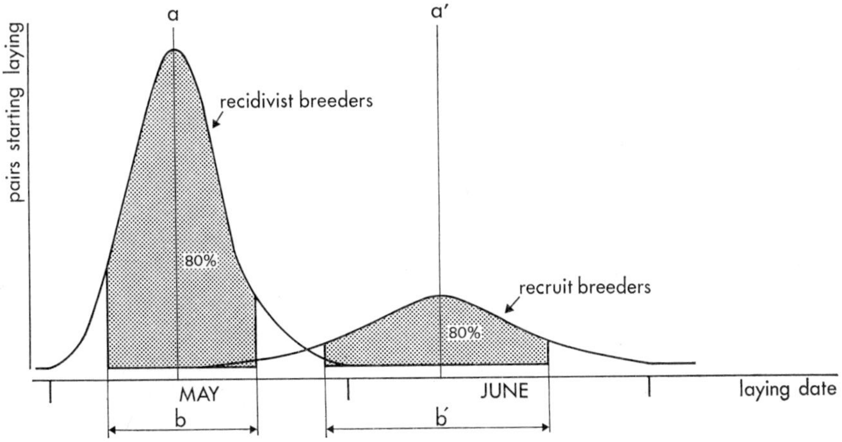

Fig. 48. Temporal distribution of egg-laying of the sandwich tern population on Griend (in outline).

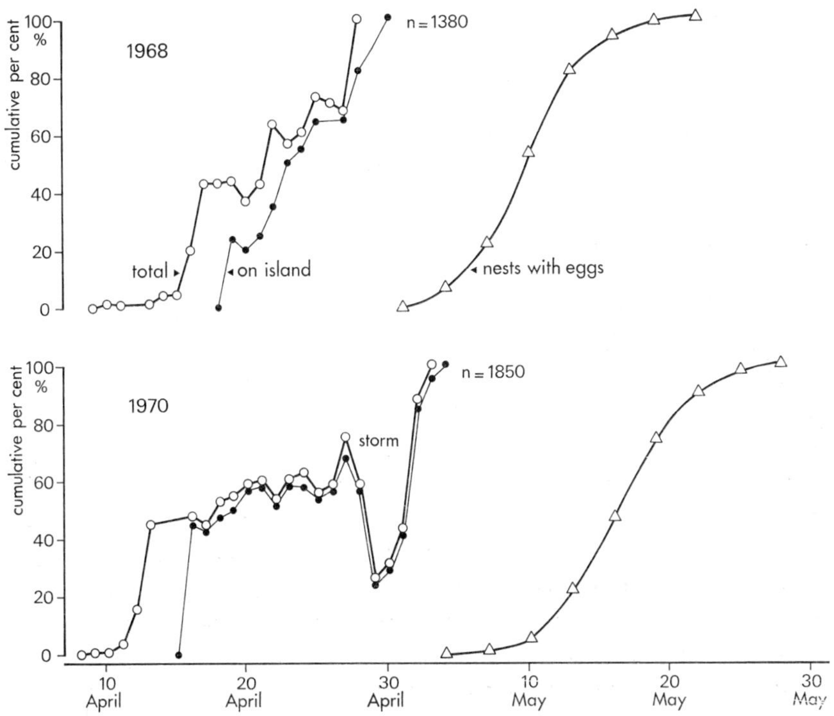

Fig. 49. Distribution of the number of sandwich terns arriving in the neighbourhood of Griend (open circles); distribution of the number of terns concentrated on the ridge of dunes on the island (black dots); and distribution of the number of pairs starting egg-laying (triangles).

to investigate to what extent differences in arrival can explain the differences in the temporal pattern of egg-laying observed.

If abiotic environmental factors through time of arrival were the only factors determining laying date, one would expect the distribution of the number of birds arriving to have a course which is parallel to the distribution of the number of pairs starting egg-laying. Figure 49 gives both graphs for the recidivist breeders in 1968 and 1970. It shows that the graphs by no means parallel each other; egg-laying took place in a shorter period than arrival. This is an indication that besides time of arrival, other factors play a role in determining island synchronization.

MARSHALL & DISNEY (1957) have experimentally shown that first-year *Quelea quelea* are sexually stimulated by the green grass that grows after heavy rainfall in tropical Tanzania. Thus, visual stimuli produced by the breeding habitat may have an important synchronizing effect on egg-laying within a bird population. In the Griend situation, in the pre-laying period the sandwich terns suddenly move to the nesting area after a period of concentration on the tidal flats. If visual stimuli from the nesting area determine egg-laying in this species and if such an influence is dependent of the length of the period of concentration, then the distribution of the number of terns present within the nesting area instead of the number arriving in the neighbourhood of the island can be expected to have a course which is parallel to the distribution of egg-laying. In Figure 49 the distribution of the daily maximum number of terns present on the ridge of dunes of the island in 1968 and 1970 has been added to the graphs on arrival and egg-laying. In 1968, after a sudden invasion within the area, the number of terns on the ridge increased proportionally to the number arriving in the vicinity of the island. In 1970, however, all terns present near the island alighted on the ridge on the 19th of April. Thereupon, in contrast to the 1968 situation, the majority of terns arriving immediately joined those already present on the ridge. In 1968 the distribution of the number of terns on the island parallels the distribution of egg-laying fairly well. In 1970, however, both graphs do not. The 1970 results thus indicate, that besides a possible influence of habitat stimuli, other factors must be involved in determining date of laying and degree of synchrony of the recidivist breeders of the island population.

In the Griend situation the amount of food available for the sandwich tern may change considerably. Sandwich terns are dependent on only a few prey-species which, under the influence of sea temperature, tend to migrate to coastal waters in spring. Besides, prey catchability strongly depends on weather circumstances which in the course of spring gradually become more favourable. For a number of species the onset of laying has been shown to be determined by the amount of food available for the female in the pre-laying period (PERRINS, 1965, 1966; LACK, 1968). In the case of the sandwich tern a sudden increase of the number of prey available thus might have an important synchronizing effect on egg-development and, consequently, on the timing of egg-laying (laying date as well as degree of synchrony) of the recidivist breeders of the population. Unfortunately, data on the availability of food in the neighbourhood of the Griend colony are insufficiently complete to test this idea.

In a number of species visual and auditory stimuli from conspecific birds have an important effect on hormone production, which in turn influences gonadal development and egg production (LEHRMAN, 1955, 1959; YOUNG, 1961). It has been suggested that in a similar way exposure to visual and auditory stimuli within dense breeding concentrations may have an influence on gonadal development and egg-laying (COULSON & WHITE, 1959; NELSON, 1966, 1970). In the pre-laying period the sandwich tern concentrates in large groups in the breeding area. During concentration the terns are subject to visual and auditory stimuli of a relatively high number of conspecifics involved in courtship displays, nesting- and aggressive activities. If, in this situation, the terns are behaviourally and physiologically affected by these stimuli, then, concentration in the pre-laying period might have an important effect on the timing of egg-laying. In this respect it is of interest that recidivist breeders arriving in the breeding area relatively late, require less time between arrival and laying than do breeding birds arriving earlier. A comparison of the graphs on arrival and egg-laying (Fig. 49) demonstrates this. In fact we have no firm proof that the order in which the terns arrived on the island was the same as the order in which they started egg-laying. However, the second influx of terns on the 23rd of April 1968 included several individuals stained with oil and a relatively high number of oil stained birds started egg-laying between 10 and 12 May. This suggests that the order of birds arriving and egg-laying was indeed the same.

Similar results have been obtained by NELSON (1966, 1970) for the North Atlantic gannet. According to NELSON: "This effect cannot be due to a specific 'year clock' causing laying between fixed dates irrespective of the pre-laying period spent in the colony, since then it could not happen, as it does, that late comers never spend longer, but always less time (and proportionately less) between return and laying. Thus a given number of weeks of high tempo behaviour stimulation had more physiological effect than the same period of lower tempo activity and resulted in a reduction of laying date differences introduced by individual variation, *i.e.* it caused a degree of synchronization." In principle I agree with NELSON, though it might be argued that the relationship between late arrival and reduction of the time spent between arrival and egg-laying in case of the sandwich tern can also be explained by a possible synchronizing effect of food becoming more and more plentiful in the course of the season.

3.2. GROUP SYNCHRONIZATION

In the sandwich tern the occupation of the nesting territory is a remarkably quick procedure: groups of birds suddenly appear on the island in daytime, individual pairs immediately occupy a territory and within a few days after appearance the first egg is laid. Different groups appear in the area with intervals of only a day up to more than a week. Within each group egg-laying is markedly synchronized.

In a number of species differences in laying date and degree of synchrony between groups of breeding birds have been explained as a consequence of differences in social stimulation received by the members of these groups (NELSON, 1970; COULSON & WHITE, 1959). If the same holds for the sandwich tern, and it seems the most likely explanation of group synchronization in this species, we can think of an effect of social stimulation on different groups of birds in two different periods; (a) in the period prior to settlement and (b) between settlement and egg-laying.

(a) On Griend from the middle of April onwards large groups of sandwich terns concentrate in the nesting area during the evening. Within these groups courtship activity is extremely high (see also I.2.1. and III.2.1.). Usually the terns gathered in only two or three large concentrations. Since these groupings were by no means identical with the much smaller breeding groups which settled on the island later in the season, differences in courtship activity between tern concentrations present on the island in the period prior to settlement cannot be simply related with differences in laying date and degree of synchrony between groups of breeding birds. However, among the birds concentrated on the island in the period prior to settlement group displays have been observed which may have had an influence on the sexual development of only a relatively small part of the large number of birds present. These displays, termed 'screaming panics' (VAN DEN ASSEM, 1954c), can be described as follows:

A screaming panic is always initiated by a particular call which sounds like a harsh 'kurreet', given by a bird (or pair) flying over the group. In reaction to this call several birds, both paired and unpaired individuals, start calling in the same way. They all adopt a very pronounced stretch posture (for description of postures see III.2.1.) and raise their head feathers conspicuously. After the flying initiator has landed it is approached from all sides by the birds responding on the ground. Several aggressive meeting ceremonies between initiator and ground birds and between the ground birds mutually then follow. Among the various postures shown the stretch posture, erect posture, circling and overt aggression predominate. In general, an assembly of screaming birds desintegrated within a few minutes. The whole procedure, however, usually caused a marked revival of display activities among its participants which might last tens of minutes and presumably even longer.

Screaming panics definitely have an important stimulating effect on courtship displays between mates. They thus might accelerate sexual development in these birds. Moreover, if the same birds repeatedly meet each other when participating together in screaming panics, this could be the mechanism through which synchronization of sexual development in groups of birds, later on leading to synchronized egg-laying, is accomplished. Unfortunately, it was not possible in the pre-laying period to watch the terns from a distance that would allow us to follow individual birds or to do other detailed observations. As a consequence we do not know whether the same birds indeed meet each other during screaming panics.

(b) The settlement on the island at different times of groups of breeding birds might also be explained if a mechanism would exist by which pairs with fully developed eggs recognize this stage in each other, and by which pairs in the same state would subsequently be induced to settle together as soon as their number has passed a threshold. As yet no evidence in support of this hypothesis is available.

4. ASSOCIATIVE NESTING OF SANDWICH TERNS WITH COMMON TERNS AND BLACK-HEADED GULLS

It has been shown that associative nesting of sandwich terns with common terns and black-headed gulls is a rule in the Netherlands (I.2.3.). Moreover, it has become clear that the sandwich tern in general benefits from nesting in the vicinity of other species. This raises the question, whether nesting in association is the result of inter-specific attraction or simply the outcome of a similar habitat preference of different species.

4.1. OCCUPATION OF THE NESTING AREA OF THE DIFFERENT SPECIES

It is well known that common terns, arctic terns and black-headed gulls tend to return to the same area year after year. This phenomenon, called site-attachment, seems poorly developed in the sandwich tern. Ring recoveries (VEEN, unpubl.) have shown that sandwich terns often change from one colony to another from year to year. Moreover, the species often occupies a completely new area for breeding or suddenly disappears from an area where it has bred for many years in succession. Figure 50 gives the distribution of sandwich tern colonies on Griend between 1966 and 1968. It demonstrates that the sandwich tern, also within the same area, may select different nesting places in consecutive years.

From many colonies it has been reported that the sandwich tern arrives later in the area than the black-headed gull (ROOTH, 1965; CULLEN, 1960a). Figure 51 gives the proportion of sandwich terns, common terns and black-headed gulls present on Griend in 1968 and 1970 in the pre-laying period. It shows that on Griend the different species arrived in the breeding area roughly at the same time. Common tern and black-headed gull occupy a nesting territory immediately after arrival on the island. The sandwich tern, however, does not occupy a definite site until the end of April or the beginning of May. Consequently, the sandwich terns selected a nesting site roughly two weeks later than common terns and black-headed gulls.

Several examples are available of sandwich terns settling in a new breeding area in the neighbourhood of common terns, arctic terns or black-headed gulls, whereas to the human eye plenty of appropriate sites were available elsewhere (CULLEN, 1960a; LIND, 1963; SALOMONSEN, 1947). It has also been reported that when common terns or black-headed gulls change their colony site within the area, the sandwich terns seem to follow (KRUUK, pers. comm.; reports of the wardens of a number of colonies in the Netherlands). Sometimes the sandwich

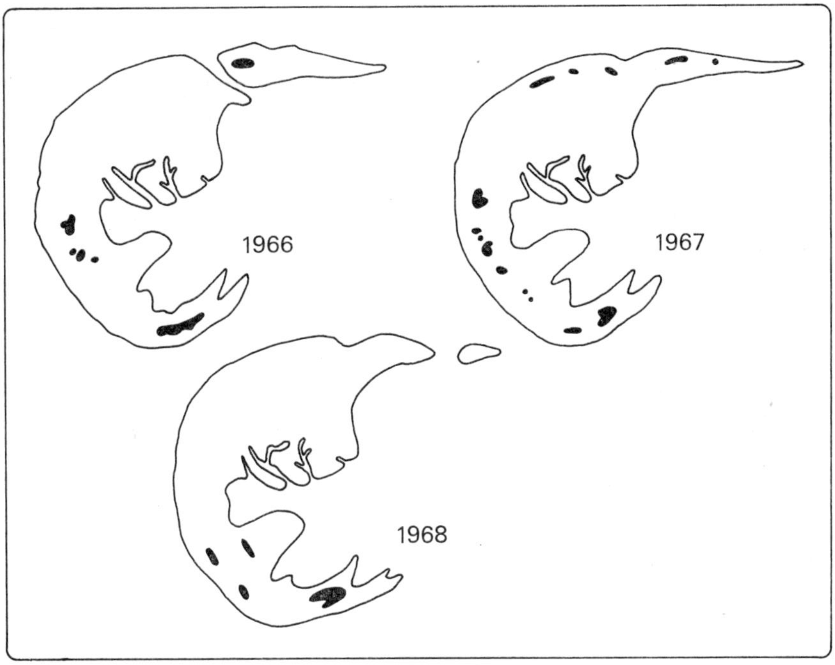

Fig. 50. Distribution of colonies of the sandwich tern on Griend in 1966, 1967 and 1968.

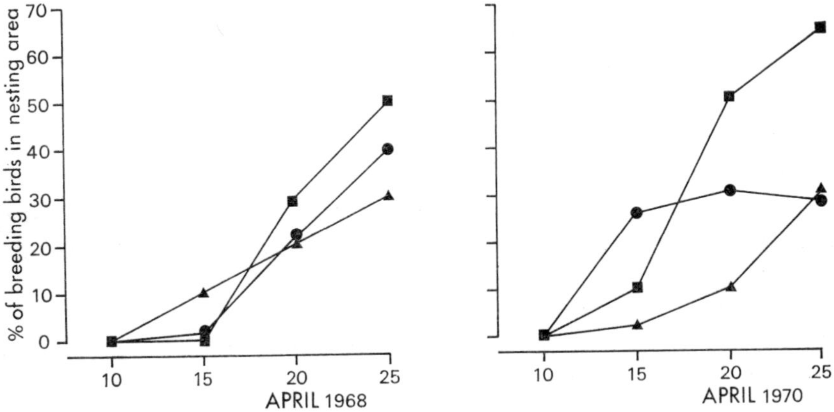

Fig. 51. Occupation of the nesting area by sandwich tern, common and arctic tern (together) and black-headed gull on Griend in 1968 and 1970.

tern is said to penetrate into an area already occupied by common terns or black-headed gulls. SALOMONSEN, for instance, gives the following picture for a colony on Hirsholmene (Denmark): 'The sandwich tern arrives later than the other species. It forces its way to the centre of the black-headed gull colonies, in the midst of the populated area, destroying the eggs and driving all birds away.' A similar description of sandwich terns settling in the middle of a colony of black-headed gulls is given by the wardens of the former colonies on De Beer and on Schouwen (Netherlands). In both colonies the phenomenon of settling inside the gullery occurred as a rule (KORFMAKER and VAN DE HOEK, pers. comm.).

The above described aspects of the behaviour of the sandwich tern, in my opinion, strongly support the idea that this species selects a colony-site under the influence of inter-specific attraction.

4.2. AN EXPERIMENT WITH BLACK-HEADED GULL DUMMIES

When I did my observations on Griend, black-headed gulls, common terns and arctic terns bred all over the island and the sandwich terns obviously had no other choice than to settle in the neighbourhood of one of these species. This made it impossible to analyse under natural circumstances the role of the presence of common terns and black-headed gulls in site-selection of sandwich terns. Therefore, an experiment has been done in 1967, 1970 and 1971 in which I tried to manipulate the distribution of sandwich terns on the island by placing black-headed gull dummies in the area in the pre-laying period.

In 1967 20 black-headed gull dummies and 20 sandwich tern dummies were simultaneously presented to study the reactions of the terns to both types of dummies. Method and results have been given in III.2.3. During this experiment the terns did not noticeably react to the gull dummies. However, we cannot conclude from these results that sandwich terns do not react to black-headed gull dummies. The interest of the terns in the sandwich tern dummies might have hidden their interest in the gull dummies. To test this, the experiments have been continued in 1970 and 1971 using the gull dummies alone.

In 1970 two different squares were pegged out. A group of 20 black-headed gull dummies was in different tests alternately placed in one of both. Observations were made from 17.00 till 20.00 hours every day between 13 April and 11 May. The results of the experiment were fully negative. Neither sandwich terns nor common terns or black-headed gulls were recorded in the squares.

Plates 22 and 23. An experiment to test the influence of size and density of a group on its attractiveness; plate 22 shows a widely spaced group and plate 23 a densely spaced group of sandwich tern dummies. (phot. U. Hirsch).

22

23

24

25

In 1971 three different squares were pegged out. Two groups of 20 black-headed gull dummies were simultaneously presented. In one group the dummies were spaced 50 centimeter apart, in the other 1-2 meters. Both groups changed positions between the squares. Observations were made from 18.00 till 20.00 hours every day between 29 April and 11 May. Once more the results were completely negative: sandwich terns, common terns and black-headed gulls were not recorded in the squares.

Summarizing we conclude that the sandwich terns were not noticeably attracted by the black-headed gull dummies.

The black-headed gull dummies were prepared in the same way as the sandwich tern dummies. Although we have no proof that both types of dummies represented their species in a similar degree, we have no arguments to accept the reverse. (The absence of a reaction to the gull dummies of living black-headed gulls may be explained by the surface features of the area in which the experiment was carried out; the area must have been unattractive for black-headed gulls.) The most reasonable explanation for the different reactions of the sandwich terns to the tern- and the gull dummies is that the stimuli presented by both types of dummies are assimilated in a different way. It may be useful to distinguish between the following alternatives:

(a) A variety of bird species develop through a learning process an image of the breeding habitat in the environment in which they grow up. Such learning processes include many details. If we assume that the sandwich tern develops an image of its breeding habitat on the place of birth, then common terns and black-headed gulls nesting in the neighbourhood might be included in the image. Since common terns and black-headed gulls have a positive influence on the breeding success of the sandwich tern, sandwich terns nesting in association with these species will raise a relatively high proportion of chicks. As a consequence, an increasing number of sandwich terns will develop a 'habitat image' in which common tern and black-headed gull are present. This might have resulted in the present situation in which the majority of sandwich terns tend to settle in an area in which common terns and black-headed gulls are present.

(b) In contrast to the above hypothesis, in the sandwich tern species specific characters of other larids might be genetically 'blue-printed'. As a consequence nesting in association with these species would take place independently of a detailed 'learning process' as suggested above.

Plate 24. Groups of sandwich terns selected a nesting site in the neighbourhood of groups of dummies previously positioned on the island. The plate illustrates that in general the distances between the nests and the dummies were remarkably short.

Plate 25. A black-headed gull dummy. The sandwich terns never reacted to this type of dummy.

In this context it is interesting to go into the historical distribution pattern of common tern, arctic tern and black-headed gull. Common tern and arctic tern can be regarded as 'authentic' coastal breeders. This, however, is not the case with the black-headed gull. According to VOOUS (1960) the black-headed gull has increased in numbers during the last century and only recently has the species settled as a breeding bird along the coast. Thus, if associative nesting with other species has evolved under the influence of the advantages of the presence of such species, it must have developed primarily in the presence of common terns and arctic terns. The black-headed gull, immigrating in the area afterwards, might have fitted the image of 'white colony-breeding larids' by which the terns were attracted. According to this hypothesis, the sandwich tern would settle in the neighbourhood of black-headed gulls, irrespective of its advantages or disadvantages.

With respect to the experiments carried out with sandwich tern and black-headed gull dummies it should be mentioned that the results obtained fit in best with the first hypothesis. The existence of a less detailed 'genetic image' of conspecifics, together with a detailed image of the black-headed gull as a result of a 'learning process' might explain the different reactions to sandwich tern and black-headed gull dummies.

SUMMARY

I. Introduction.

The main part of this study is an attempt to investigate whether in the sandwich tern colonial nesting enhances reproductive success, and if so how this comes about (chapter II). Besides, attention is paid to the behavioural causation of colonial nesting (chapter III). Considering the various ways in which colonial nesting might influence success, a study of function would ideally require a comprehensive study of all environmental factors. Although the relationship between nest distribution and several environmental factors will be touched upon in this study, my concern has been primarily with predation. This choice has been inspired by the work done by TINBERGEN and co-workers studying nest distribution in gulls.

According to TINBERGEN and co-workers, the spatial pattern of nest distribution in a black-headed gull colony should be regarded as a consequence of two opposing tendencies, 'clustering' and 'spacing-out', which both have an anti-predator function: Nest spacing assists concealment of the nests which in turn lowers the risk that eggs and chicks are discovered by predators, while clustering has the advantage that a predator can be simultaneously attacked by a large number of breeding birds.

Similarities in behaviour and nest distribution of Larid species suggest that the majority of tern and gull species of the northern hemisphere defend their brood against predators in a basically similar way. However, there are some exceptions among which is the sandwich tern. Sandwich terns nest at pecking distance of each other, the nesting area is rather conspicuous and the birds are not very inclined to attack predators. This aberrant behaviour cannot be explained by the absence of predators for a variety of predator species has been recorded from sandwich tern colonies.

In the case of the sandwich tern there is another point of interest which might be related to predation. Sandwich terns tend to nest in association with common terns, arctic terns and black-headed gulls. It has been suggested that the sandwich tern might gain protection against predators through the aggressiveness of these neighbours. However, it has also been noted that black-headed gulls take a heavy toll from the terns by stealing eggs, chicks and fish.

II. Functional aspects of nest distribution.

A large number of sandwich tern nests has been under continuous observation. For each of these nests reproductive success has been determined. Clutch size, group synchronization, nest position (density, inter-nest distance and central vs peripheral position), group size, colony size and island synchronization were related independently of each other to success. The influence of age differences and environmental factors correlated with the season were excluded from the analysis as much as possible. Except for group size and central vs peripheral nest position, all aspects mentioned above were correlated with hatching and/or fledging success.

The sandwich tern population at Griend consists of two groups of breeding birds, the recidivists and the recruits, which are different with respect to their date of egg-laying. The recruits are strikingly less successful than the recidivists. This can be partly explained by differences in clutch size, group synchronization, nest position and colony size. However, other factors such as environmental influences correlated with the season must play a role as well.

Data on mortality agents collected between 1966 and 1972 showed that mortality of eggs and chicks was caused by predation (eggs 14.3%, chicks 11.7%), high tides (eggs 3.8%), hatching failure (eggs 6.9%), food shortage, diseases and physical defects (together 33.0% of chicks).

Black-headed gull, herring gull, great black-backed gull, oystercatcher, arctic skua, short-eared owl and several raptor species have been recorded in the sandwich tern colonies on Griend. Black-headed gull, herring gull and great black-backed gull were the only species preying upon the brood of the sandwich tern in any important way. Only a small proportion of the total number of gulls present on the island took part in predation on the brood of the terns. In the case of the black-headed gull the number of predators tended to increase in the course of the season. The number of herring- and great black-backed gulls preying upon the sandwich tern was limited by inter-specific aggression. Black-headed gulls have been recorded to rob eggs and chicks; herring- and great black-backed gulls mainly robbed chicks. A relatively high proportion of the eggs and chicks robbed by black-headed gulls should be regarded as potentially non-viable. Herring gulls and great black-backed gulls, however, took potentially viable chicks from the population. Considering two years of observation together, the influence of predation by black-headed gulls and herring gulls (great black-backed gulls included) on the breeding success of the sandwich tern was roughly the same.

Piracy by black-headed gulls on the food of the sandwich tern influences chick mortality when food gets short in a long lasting period of stormy weather. However, such periods are rare in the hatching season of the terns.

With respect to predation and food piracy the situation on Griend can be regarded as representative for most other colonies in the Netherlands.

To a predator which endangers the brood only, sandwich terns react with sitting down on the nest, threatening from a sitting position. To a predator which is a danger to the brood and to the adults, the terns react with flying up. In reaction to a predator which endangers only the adults, flying up from the nests, followed by escape behaviour is shown. Sometimes the predator is attacked from the air, but usually the attack frequency and intensity is low. The tendency of the terns to attack a predator strongly increases after they have left the nest shortly after hatching.

Common tern and black-headed gull tend to fly up in reaction to nearly all predator species. Predators of the brood are vigorously attacked.

The anti-predator behaviour of the sandwich tern defends the brood effectively against black-headed gulls, oystercatchers, herring gulls and great black-backed gulls. It gives little protection to the brood when avian or mammalian ground predators are involved which endanger both adults and brood.

The behaviour of predators reacting to the attacks of common tern, black-headed gull and sandwich tern mainly depends on attack frequency and intensity. Several predator species are successfully driven off from the nesting area by the combined attacks of the different breeding birds. The protective influence for the sandwich tern of common terns and black-headed gulls nesting in the neighbourhood has been substantiated in the case of predation by black-headed gulls, herring gulls and great black-backed gulls.

The discussion deals with (a) the mechanisms through which the various aspects of nest distribution might reduce predation, (b) the advantages and disadvantages of associative nesting with another species and (c) the selective advantage of the sandwich tern nest distribution pattern with respect to some environmental factors other than predation.

In the Griend situation the advantage of nesting together with black-headed gulls very likely outweighs its disadvantages. In colonies where the sandwich tern has to cope with mammalian ground predators, associative nesting has a number of advantages in addition to the ones established for the Griend situation. Therefore, it is concluded that associative nesting with common terns and black-headed gulls must be advantageous for the sandwich tern in general.

It is suggested that the nest distribution pattern of the sandwich tern and behaviour aspects related to it have evolved under the pressure of at least three different factors: food, predation and tidal movements. Exploitation of food resources may have selected for colonial nesting. Nesting aggregations will attract predators. This may have led to a selection pressure for choosing a habitat in which mammalian ground predators are scarce. However, the habitat chosen introduced the danger of high tides, which in turn may have selected for special behaviour adaptations, such as little or no attachment to the site of the previous year, selection of relatively high nesting places and the possibility for chicks to move within the area. The remarkably high nest density of the sandwich tern may have evolved as a consequence of the need for breeding synchronization. However, a shortage of nesting places safe from high tides may have selected for dense nesting as well.

III. Causal aspects of nest distribution.

This section deals with the question of how behaviour patterns leads to the spatial and temporal nest distribution pattern and to associative nesting in the sandwich tern.

Throughout the pre-laying period the terns of the Griend population go through a number of successive stages which are characterised by an increase of group density and an increase of nesting and aggressive activities. House-hunting individuals select a nesting site at very short distance of nesting conspecifics (clustering), but territorial aggression prevents them from coming too close (spacing out). There is a special relationship between incubating and non-incubating individuals; the sitting position of incubating birds seems to inhibit aggression of non-incubating individuals. Differences in minimum inter-nest distance between peak -and tail breeders of a group are explained by behaviour differences which are related with differences in phase of breeding of neighbouring pairs at the moment of nest site selection. Experiments with sandwich tern dummies have thrown some light on the influence of size and density on group attractiveness. The nature of the reactions of the terns to the dummies is investigated.

The possible ways in which synchronization of egg-laying of the island population and of separate groups may be effected are discussed.

Evidence is put forward which support the idea that associative nesting of sandwich terns with common terns and black-headed gulls is the result of interspecific attraction. An attempt to manipulate the distribution of sandwich terns on the island by placing black-headed gull dummies in the area was unsuccessful. This failure of the black-headed gull dummies, in sharp contrast to the effectiveness of the similarly made sandwich tern dummies, leads to the supposition that the terns make higher demands upon conformity with the living bird for the gull dummies than for the tern dummies. This is explained by differences in the degree to which the recognition of conspecifics and other species is based on detailed learning.

REFERENCES

ASHMOLE, N. P. (1962). The Black Noddy *Anous tenuirostris* on Ascension Island. Part I. General biology. — Ibis 103, p. 235-273.

—— (1963a). The biology of the Wideawake or Sooty tern *Sterna fuscata* on Ascension Island. — Ibis 103, p. 297-364.

—— (1963b). The regulation of numbers of tropical oceanic birds. — Ibis 103, p. 458-473.

ASSEM, J. van den (1954). Waarnemingen over het gedrag van de Grote stern. — De levende Natuur 57, p. 1-9.

—— (1954b). Kokmeeuwen als belagers van Grote sterns. — De Levende Natuur 57, p. 141-145.

—— (1954c). Reproductive behaviour in the Sandwich tern, *Sterna s. sandvicensis*. — Unpublished report, Leiden University.

AUSTIN, O. L. (1945). The role of longevity in successful breeding of the Common Tern. — Bird Banding 16, 21-28.

BROEKHUIZEN, S., & SMIDT, A. de (1965). Het verloop van het broedseizoen 1964. op Griend en een verzicht van alle waargenomen vogels. — De Levende Natuur 68, p. 25-34.

BROUWER, G. A., DIEREN, J. W. van, FEEKES, W., HARMSEN, G. W., HOUTEN, J. G. ten, KABOS, W. J., MAZURE, J. P., SCHEYGROND, A., TESCH, P., & WERFF, A. van der (1950). Griend het vogeleiland in de waddenzee. — 's Gravenhage, Martinus Nijhoff.

COMMISIE voor de Nederlandse Avifauna. (1962). Avifauna van Nederland. — Ardea 50, p. 16.

COULSON, J. C., & WHITE, E. (1958). The effect of age on the breeding biology of the Kittiwake, *Rissa tridactyla*. — Ibis 100, p. 40-51.

—— & —— (1959). The effect of age and density of breeding birds on the time of breeding of the Kittiwake *Rissi tridactyla*. — Ibis 101, p. 496-497.

—— (1966). The influence of the pair-bond and age on the breeding biology of the Kittiwake gull *Rissa tridactyla*. — J. Anim. Ecol. 35, p. 269-279.

CROOK, J. H. (1965). The adaptive significance of avian social organisation. — Symp. Zool. Soc. Lond. 14, p. 181-218.

CROZE, H. (1970). Searching Image in Carrion Crows. — Parey Verlag, Berlin.

CULLEN, E. (1957). Adaptations in the Kittiwake to cliff nesting. — Ibis 99, p. 275-302.

CULLEN, J. M. (1960a). Some adaptations in the nesting behaviour of terns. — Proc. XIIth Int. Orn. Congr. Helsinki 1958, p. 153-157.

—— (1960b). The aerial display of the Arctic tern and other species. — Ardea 48, p. 1-37.

DIRCKSEN, R. (1932). Die Biologie des Austernfischers, der Brandseeschwalbe und der Küstenseeschwalbe nach Beobachtungen und Untersuchungen auf Norderoog. — J. f. Ornith. 80, p. 427-521.

HAWKSLEY, O. (1957). Ecology of a breeding population of Arctic Terns. — Bird Banding 29, p. 57-92.

KOEMAN, J. H., OSKAMP, A. A. G., VEEN, J., BROUWER, E., ROOTH, J., ZWART, P., BROEK, E. van den & GENDEREN, H. van (1967). Insecticides as a factor in the mortality of the Sandwich tern (*Sterna sandvicensis*). A preliminary communication. — Med. Rijksfac. Landbouwwet. Gent 32, p. 841-854.

——, Veen, J., Brouwer, E., Huisman de Brouwer, L. & Koolen, J. H. (1968). Residues of chlorinated hydrocarbon incesticides in the North Sea environment. — Helgoländer Wiss. Meeresunt. 17, p. 375-380.

Krebs, J. R. (1974). Colonial nesting and social feeding as strategies for exploiting food resources in the Great blue heron (*Ardea herodias*). — Behaviour 51, p. 99-131.

Kruuk, H. (1964). Predators and anti-predator behaviour in the Black-headed gull (*Larus ridibundus*). — Behaviour Suppl. 11, p. 1-130.

Lack, D. (1947). The significance of clutch size. — Ibis 89, p. 302-352.

—— (1948). The significance of clutch size. — Ibis 90, p. 25-45.

—— (1954). The natural regulation of animal numbers. — Oxford Univ. Press.

—— (1966). Population studies of birds. — Clarendon Press, Oxford.

—— (1968). Ecological adaptations for breeding in birds. — Methuen, London.

Lehrman, D. S. (1955). The physiological basis of parental feeding behaviour in the Ring dove. — Behaviour 7, p. 241-286.

—— (1959). Hormonal response to external stimuli in birds. — Ibis 101, p. 478-496.

Lind, H. (1963). Nogle sociale reactioner hos terner. (Notes on the social behaviour in Terns). — Dansk Orn. Foren. Tidsskr. 57, p. 155-176.

Marshall, A. J. & Disney, H. J. de S. (1957). Experimental induction of the breeding season in a xerophilous bird. — Nature 180, p. 647.

Meyer-Waarden, P. F. & Tiews, K. (1965). Der Beifang in den Fängen der Deutschen Garnelenfischerei in den Jahren 1954-1960. — Ber. Dt. Wiss. Komm. Meeresforsch. XVIII, p. 13-78.

Müller, H. (1959). Zugverhältnisse der europäischen Brandseeschwalben. — Die Vogelwarte 20, p. 91-115.

Muus, B. J. (1966). Zeevissengids. — Elsevier, Amsterdam.

Nelson, J. B. (1966). The breeding biology of the Gannet, *Sula bassana* on the Bass Rock, Scotland. — Ibis 108, p. 584-626.

—— (1967). Colonial and cliff nesting in the Gannet. — Ardea 55, p. 60-70.

—— (1970). The relationship between behaviour and ecology in the Sulidae with reference to other sea birds. — Oceanogr. Mar. Biol. An., Rev. 8, p. 501-574.

Patterson, I. J. (1965). Timing and spacing of broods in the Black-headed Gull. — Ibis 107, p. 433-453.

Perrins, C. M. (1965). Population fluctuations and clutch-size in the Great Tit, *Parus major* L. — J. Anim. Ecol. 34, p. 601-647.

—— (1966). Survival of young Manx Shearwaters *Puffinus puffinus* in relation to their presumed date of hatching. — Ibis 108, p. 132-135.

Postuma, K. J., Zijlstra J. J. & Das, N. (1965). On the immature herring of the North Sea. — J. du Conseil Int. Explor. Mer 29, p. 256-276.

Rooth, J. (1958). Relations between Black-headed Gulls (*Larus ridibundus*) and terns (*Sterna* spec.) in the Netherlands. — IIIrd Bull. Int. Comm. Bird Preservation, p. 117-119.

—— (1965). Over sterns and kaapmeeuwen.—De Levende Natuur 68, p. 265-275.

Rutten, M. (1931). Het roven van eieren en jongen van sterns (*Sterna* spec.) door scholeksters op de Beer. — Ardea 20, p. 176.

Salomonsen, F. (1947). Maagekolonierne paa Hirsholmene. — Dansk Orn. Foren. Tidsskr. 41, p. 147-186.

Smidt, A. de & Broekhuizen, S. (1964). Griend, 1964, waarnemingen over het verloop van het broedseizoen en onderzoek naar de relatie tussen kapmeeuwen en grote sterns. — Doctoraal verslag, Utrecht.

Tiews, K. (1971). Weitere Ergebnisse von Langzeitbeobachtungen über das Auftreten von Beifangfischen und -krebsen in den Fängen der deutschen Garnelenfischerei. (1961-1967). — Arch. Fischereiwiss. 22, p. 214-255.

Tinbergen, N. (1959). Comparative studies of the behaviour of gulls: A progress Report. — Behaviour 15, p. 1-70.

——, Broekhuysen, G. H., Feekes, F., Houghton, J. C. W., Kruuk, H. & Szulc, E. (1962). Egg-shell removal by the Black-headed gull *Larus ridibundus* a behavioural component of camouflage. — Behaviour 19, p. 74-117.

——, Impekoven, M., & Franck, D. (1967). An experiment on spacing-out as a defence against predation. — Behaviour 28, p. 307-321.

Veen, J. (1967). Gechloreerde koolwaterstoffen in eieren en weefsels van de Grote Stern (*Sterna s. sandvicensis* Lath.) in Nederland. — Doctoraal verslag, Inst. voor Vet. Farm. en Toxicologie. Utrecht.

Voous, K. H. (1960). Atlas der Europese vogels. — Elsevier, Amsterdam.

Ward, P. & Zahavi, A. (1973). The importance of certain assemblages of birds as 'information centres' for food finding. — Ibis 115, p. 517-534.

Young, W. C. (1961). (Editor), Sex and Internal Secretions, Vols I and II. — Ballière, Tindall and Cox, London.

RÉSUMÉ

I. Introduction.

La majeure partie de cette étude est consacrée à une tentative d'établir si la nidification en colonies contribue au succès reproductif du Sterne caugek et, dans l'affirmative, par quelle voie ce résultat est obtenu (chapitre II). En outre, les causes de la nidification en colonies seront considérées du point de vue du comportement (chapitre III). Lorsqu'on examine les différentes façons dont la nidification en colonies peut influencer le résultat reproductif, une recherche fonctionelle exigerait, en principe, une investigation exhaustive de tous les facteurs extérieurs. Bien que la relation entre la disposition des nids et les divers facteurs extérieurs sera abordée, mon attention s'est portée principalement sur la "prédation". Ce choix m'a été inspiré par les recherches de TINBERGEN et de ses collaborateurs sur la disposition des nids des Goélands et des Mouettes.

Selon TINBERGEN et ses collaborateurs, la configuration spatiale de la disposition des nids dans une colonie de Mouettes rieuses doit être vue comme l'effet de deux tendances opposées, notamment celles de rapprochement et de dispersion, qui toutes deux jouent un rôle de protection contre les prédateurs: les nids espacés se remarquent moins, ce qui réduit le risque de découverte des œufs et des poussins, tandis que le rapprochement des nids présente l'avantage qu'un prédateur peut être attaqué simultanément par un nombre élevé d'oiseaux couvants.

Les ressemblances en comportement ainsi qu'en disposition des nids parmi les espèces de Laridés permettent de penser que les Mouettes et les Goélands de l'hémisphère du Nord défendent leur progéniture contre les espèces rapaces d'une manière fondamentalement similaire. Il y a cependant des espèces qui font exception et le Sterne caugek en est une. Les Sternes caugek font leurs nids tellement près les uns des autres, qu'ils peuvent se toucher mutuellement à coups de becs. Dès lors, la zone où les nids se trouvent peut-elle se remarquer sans grande difficulté. Par ailleurs, les Sternes caugek n'ont pas tendance à attaquer les prédateurs. On ne peut expliquer cette différence de comportement par l'absence de prédateurs, car plusieurs espèces rapaces ont été signalées fréquentant les colonies des Sternes caugek.

Le Sterne caugek est doué d'une autre particularité qui pourrait être liée à la prédation. Les Sternes caugek préfèrent nicher ensemble avec des Sternes pierregarin, des Sternes arctiques et des Mouettes rieuses. Il a été supposé que ce voisinage pouvait protéger les Sterne caugek contre les prédateurs grace à l'agressivité de tels voisins. Mais il a été observé que les Mouettes rieuses exigent un lourd péage de la part des Sternes caugek en dérobant leurs œufs, leurs béjaunes et leurs poissons.

II. Aspects fonctionnels de la disposition des nids.

Un grand nombre de nids de Sternes caugek a été soumis à une observation continue. Pour chaque nid, le succès reproductif a été déterminé. Ce succès a été examiné, quant à son rapport, avec chacun des facteurs suivants: le nombre d'oeufs, la synchronisation du groupe, la position du nid (la densité, la distance entre les nids, la position centrale ou périphérique), le nombre de nichées du groupe ainsi que de la colonie, et le degré de synchronisation de l'île. L'influence de la diffèrénce d'âge ainsi que celle des facteurs extérieurs saisonniers fut aussi bien que possible neutralisée dans cette analyse. Tous les aspects nommés précédemment, à l'exception du nombre de nichées du groupe et de la position centrale ou periphérique, ont été examinés quant à leur corrélation avec le nombre de béjaunes réussissant à sortir de l'œuf et/ou à quitter le nid.

La population des Sternes caugek à Griend se compose de deux groupes d'oiseaux couvants: les "récidivistes" (c'est-à-dire ceux qui avaient déjà participé aux activités de reproduction au cours d'une saison antérieure) et les "recrues" (les oiseaux qui y prennent part pour la première fois). Ces deux catégories diffèrent quant à leur date de ponte. Le succès reproductif des recrues est beaucoup moins élevé que celui des récidivistes. Ceci s'explique en partie par leurs différences en nombre d'œufs, en synchronisation de groupe, en position de nids et en nombre de nichées dans la colonie. Cependant d'autres facteurs, telles les influences saisonnières jouent probablement aussi un rôle.

Les causes de mortalité en œufs et poussins enregistrées entre 1966 et 1972 se répartissent comme suit: prédation (œufs 14.3%; poussins 11.7%), marées submergeantes (œufs 3.8%), l'échec d'éclosion (œufs 6.9%), manque de ressources nutritives, maladies et déficiences physiques (ensemble 33.0% des béjaunes).

Le Goéland marin, la Mouette rieuse, le Goéland argenté, l'Huîtrier pie, le Labbe parasite, le Hibou des marais et plusieurs autres espèses rapaces ont été observées dans les colonies de Sternes caugek à Griend. Seuls la Mouette rieuse, le Goéland argenté et le Goéland marin furent responsables d'une perte considérable d'oeufs et de poussins de Sternes caugek. De l'ensemble des Goélands habitant l'île, ce ne fut qu'une partie mineure qui participa à la prédation de la progéniture des Sternes. Le nombre de Mouettes rieuses rapaces tendait à s'augmenter au cours de la saison. Le nombre de Goélands marins et argentés, qui infligeaient cette perte aux Sternes, fut limité par l'agression entre espèces. On a vu piller les Mouettes rieuses des œufs, des poussins et des poissons. Une proportion relativement élevée d'œufs et de poussins, volés par les Mouettes rieuses, peut-être considerée comme potentiellement non viable. En revanche, les poussins dérobés par les Goélands marins et argentés furent potentiellement viables. En combinant les données de deux années d'observation, l'influence de prédation sur le succès reproductif du Sterne caugek par la Mouette rieuse est presque pareil à l'influence de prédation par les Goélands. En ce qui concerne la prédation, nous pouvons voir la situation à Griend comme représentative pour la plupart des autres colonies aux Pays-Bas.

A l'approche d'un rapace, dangereux uniquement pour la nichée, un Sterne caugek se tient sur son nid, menaçant en position assise. Dans le cas où le rapace menace à la fois la nichée et l'adulte, le Sterne s'envole du nid. Devant un rapace qui ne ménace que l'adulte, ce dernier s'envole et s'enfuit. Il arrive aussi que le Sterne attaque le rapace en volant, mais ces attaques sont peu fréquentes et peu vigoureuses. Lorsque les Sternes ont quitté le nid accompagnés de leurs petits, quelques jours après l'éclosion des œufs, ils sont nettement plus agressifs vis-à-vis du prédateur.

A l'approche de toute espèce de rapine, le Sterne pierregarin et la Mouette rieuse s'envolent. Ils combattent avec vigeur tout animal qui menace leur nichée.

Le comportement anti-prédateur du Sterne caugek protège la nichée de manière efficace contre la Mouette rieuse, l'Huîtrier pie et les Goélands argenté et marin. Lorsque le rapace, qu'il soit oiseau ou mammifère, menace à la fois l'adulte et la nichée, cette dernière n'est guère protégée par un tel comportement.

La réaction des prédateurs aux attaques de Sternes pierregarin, de Mouettes rieuses et de Sternes caugek, dépend en grande partie de la fréquence et de l'intensité de ces attaques. L'attaque combinée des diverses espèces d'oiseaux couvants réussit effectivement à expulser de la région des nids plus d'une espèce rapace.

De solides éléments de preuve ont été apportés à la thèse que, par rapport à la prédation par le Goéland argenté, le Goéland marin et la Mouette rieuse, la proximité des nids de Sternes pierregarrin et de Mouettes rieuses a été bénéfique à la protection du Sterne caugek.

Dans la discussion sont exposés a) le mécanisme par lequel les divers aspects de la disposition des nids pourraient limiter la prédation, b) les avantages et désavantages de la nidification en association avec une autre espèce et c) l'avantage sélectif de la configuration de la disposition des nids des Sternes caugek en égard à d'autres facteurs extérieurs que la prédation.

On peut penser que dans une situation semblable à celle de Griend, les avantages et les désavantages d'une nidification associée avec les Mouettes rieuses, se compensent. Dans les colonies où le Sterne caugek doit combattre des prédateurs mammifères, la nidification associée présente un nombre d'avantages additionnels à ceux constatés à Griend. Dès lors, on peut conclure que la nidification associée avec le Sterne pierregarin et la Mouette rieuse est avantageuse en général.

La configuration de la disposition des nids du Sterne caugek et les aspects du comportement liés à celle-ci ont probablement évolué sous l'influence d'au moins trois facteurs différents: les ressources nutritives, la prédation et les mouvements de la marée. L'exploitation des ressources nutritives pourrait avoir poussé à la nidification en colonies. Le rapprochement des nids attirera des prédateurs. Ceci aurait pu entraîné une sélection dans le sens du choix d'un habitat exempt de prédateurs mammifères. Mais ce choix-là introduit le danger des marées hautes qui, à son tour, peut avoir conduit à des comportements spécialement adaptés, telle une fidélité de lieu moindre ou nulle, une sélection de lieux relativement élevés et la possibilité pour les poussins de se déplacer dans la zone de nidification.

III. La disposition des nids et ses causes.

Cette section concerne les caractéristiques du comportement qui conduisent à la configuration spatiale et temporelle de la disposition des nids et à la nidification associée du Sterne caugek.

Pendant la période précédant la ponte, les Sternes de la population de Griend passent par un nombre de phases qui se caractérisent par un accroissement de la densité du groupe et des activités de nidification et d'agression. A la recherche d'une résidence, les oiseaux sélectionnent un lieu de nidification très proche des nids de leurs congénère (rapprochement), mais l'agression territoriale les empêche de se rapprocher trop (dispersion). Entre les individus qui ont déjà entamé la couvaison et ceux qui ne l'ont pas encore entamée, s'établit un type de relation particulier: la position assise des oiseaux couvants semble freiner l'agression de ceux qui ne couvent pas encore. Les différences de distance minimum entre les nids de "peak and tail breeders" d'un groupe s'expliquent par les différences dans le comportement agressif, lesquelles sont elles-mêmes corrélatives avec les différences entre les phases de couvaison dans lesquelles les couples se trouvent au moment où ils choisissent le lieu de nidification. L'influence du nombre de nichées et de leur densité sur l'attractivité du groupe a été examinée expérimentalement à l'aide de Sternes caugek artificiels. Les réactions des Sternes envers les Sternes artificiels ont été étudiées.

Les façons dont la ponte se synchronise sur l'île toute entière et séparément dans les groupes qui l'habitent, sont analysées.

Des éléments de preuve sont avancés pour appuyer l'opinion suivant laquelle la nidification associée des Sternes caugek avec les Sternes pierregarin et les Mouettes rieuses est le résultat d'une attraction interspécifique. La tentative de manipuler la dispersion des Sternes caugek dans l'île, en plaçant des Mouettes rieuses artificielles dans le terrain, a échoué. Cet échec, contrastant fortement avec le succès remporté par des Sternes artificiels fabriqués de façon identique, fait supposer que les Sternes exigent que les Mouettes artificielles ressemblent plus au modèle naturel que les Sternes artificiels. Pour expliquer ceci, on suppose que l'apprentissage de détails intervient d'un moindre degré dans la reconnaissance de congénères que dans celle d'autres espèces.

SAMENVATTING

Inleiding.

Nestelende vogels kunnen zich op verschillende manieren over hun broedgebied verspreiden. Er zijn soorten, zoals de meeste zangvogels, die een tamelijk gelijkmatige verspreiding te zien geven, terwijl weer andere soorten, waaronder de meeste meeuwen en sterns, binnen het gekozen broedgebied geconcentreerd zijn in kolonies. Binnen elk der genoemde categorieën geldt dat de verspreiding der nesten sterk soortspecifiek is.

Gedurende de laatste jaren heeft een toenemend aantal biologen zich beziggehouden met de vraag in hoeverre de manier van verspreiding van de nesten binnen de kolonie van een bepaalde vogelsoort gezien kan worden als een aanpassing aan de omgeving. Zij hebben daartoe onderzocht of deze verspreiding bijdraagt tot het voortplantingssucces en zo ja, hoe zo'n bijdrage dan tot stand komt. De behaalde resultaten hebben onder meer aangetoond dat het broeden in kolonies de volgende voordelen kan hebben: Het kan de paarvorming vergemakkelijken; kan via sociale interacties leiden tot synchronizatie van het broedproces; kan uitwisseling van informatie over rijke voedselbronnen tot gevolg hebben en een gemeenschappelijke verdediging tegen roofvijanden mogelijk maken. Naast voordelen kan het nestelen in kolonies ook nadelen met zich meebrengen: Het heeft een sterke voedsel- en plaatsconcurrentie tot gevolg, vergemakkelijkt de verspreiding van ziekten en parasieten en maakt de broedende vogels door hun grote aantal opvallend voor roofvijanden.

Het hier gepresenteerde onderzoek gaat in op de vraag in hoeverre bij de grote stern het broeden in kolonies betekenis heeft voor het voortplantingssucces. In zijn meest ideale vorm zouden alle omgevingsfactoren die het voortplantingssucces kunnen beïnvloeden bij een dergelijk onderzoek betrokken moeten worden. Hoewel in mijn onderzoek verscheidene omgevingsfactoren zijn bekeken, heb ik mij in hoofdzaak bezig gehouden met de mogelijke betekenis van het broeden in kolonie als verdedigingsmiddel tegen roofvijanden. Deze keuze is vooral geïnspireerd door het werk van TINBERGEN en zijn medewerkers die het koloniebroedgedrag van de kokmeeuwen bestudeerden. Verscheidene jaren geleden werd door TINBERGEN opgemerkt dat het ruimtelijke verspreidingspatroon van de nesten in een kokmeeuwkolonie niet willekeurig tot stand komt. Enerzijds tonen deze meeuwen een sterke neiging om dicht bijeen te broeden, anderzijds bewaren zij toch ook altijd enige afstand. Met behulp van een experimentele aanpak kon worden aangetoond dat 'bijeenbroeden' het grote voordeel heeft dat een roofvijand door een groot aantal broedvogels gelijktijdig kan worden aangevallen en verjaagd. Het bewaren van 'enige afstand' bleek van positieve betekenis voor de camouflage van het legsel. Volgens TINBERGEN en zijn medewerkers ontstaat het ruimtelijk verspreidingspatroon van de nesten binnen een kokmeeuwkolonie door de tegengestelde krachten 'bijeenzitten' en 'afstand bewaren' en hebben beide krachten een functie als verdedigingsmiddel tegen roofvijanden.

Overeenkomsten in het verspreidingspatroon van de nesten in kolonies van kokmeeuwen, mantelmeeuwen, zilvermeeuwen, stormmeeuwen, noordse sterns, visdieven, Dougall's sterns en dwergsterns doen vermoeden dat de meeste soorten meeuwen en sterns zich in principe op dezelfde manier tegen roofvijanden verdedigen: Zij lijken allen te profiteren van de voordelen die 'bijeenzitten' en 'afstand bewaren' bieden tegen roofvijanden. Er zijn echter enkele uitzonderingen: Bij de drieteenmeeuw liggen de nesten veelal tegen elkaar aan. Dit nogal afwijkende gedrag kan echter worden verklaard uit het feit dat er in het broedgebied van de drieteenmeeuw nauwelijks roofvijanden aanwezig zijn. Ook bij de grote stern liggen de nesten opvallend dicht bijeen. Hier kan dit echter niet verklaard worden uit de afwezigheid van roofvijanden want in kolonies van de grote stern zijn vele soorten rovers waargenomen. Met betrekking tot de vraag hoe grote sterns zich tegen roofvijanden verdedigen valt op dat grote sterns de neiging hebben om hun kolonies te vestigen in de omgeving van kokmeeuwen, visdieven en noordse sterns. Verschillende onderzoekers hebben gesuggereerd dat de grote stern zou mee profiteren van de actie die deze nogal agressieve buren ondernemen tegen gemeenschappelijke roofvijanden. De kokmeeuw is echter zelf een belangrijke roofvijand van de grote stern want kokmeeuwen stelen eieren, jongen en visjes. Het lijkt er dus op dat de grote stern de neiging heeft om zich te vestigen in de omgeving van een andere soort die bescherming biedt tegen roofvijanden, maar die tegelijkertijd zelf als roofvijand optreedt.

Samenvattend concluderen wij dat het verspreidingspatroon van de nesten in een kolonie van grote sterns afwijkt van dat van de meeste andere sterns en meeuwen. Juist de verschilpunten hebben betrekking op gedragsaspecten die bij de andere soorten geïnterpreteerd zijn als antipredatiegedrag. De wijze waarop grote sterns in kolonies broeden, doet de vraag rijzen hoe deze soort zich tegen roofvijanden verdedigt en waarom deze vogels zoveel dichter bijeen broeden dan andere meeuwachtigen. Dit proefschrift bestaat uit drie gedeelten. Deel I geeft bijzonderheden over een aantal algemene aspecten van het gedrag en van de oecologie van de grote stern die van belang zijn als achtergrondinformatie. Achtereenvolgens worden vier aspecten behandeld, nl. de opeenvolging van broedhandelingen, de koloniestructuur, de nestplaatskeuze en het voedsel van de grote stern. In Deel II wordt gezocht naar een antwoord op de volgende vragen: Levert het verspreidingspatroon van de nesten in een kolonie van grote sterns een bijdrage tot het voortplantingssucces? Welke omgevingsfactoren zijn hierbij betrokken? In hoeverre kan het broeden in kolonies bij de grote stern gezien worden als antipredatiegedrag? In Deel III wordt nagegaan welke gedragspatronen leiden tot het specifieke verspreidingspatroon van de nesten in de kolonie van de grote stern.

I. Algemene aspecten.

In de periode die aan het leggen van de eieren voorafgaat, concentreren de grote sterns zich aanvankelijk alleen 's avonds en 's nachts in het broedgebied. In het begin tonen zij nog geen duidelijke binding met een bepaalde plaats daarbinnen. Vanaf begin mei verschijnen de sterns plotseling overdag in groepen op het eiland. Enkele dagen daarna begint de meerderheid van de paren binnen zo'n groepje met eileggen. Gebaseerd op de verspreiding van de nesten in de ruimte en de vestiging in de tijd zijn achtereenvolgens onderscheiden: De eilandpopulatie bestaande uit een aantal kolonies, de laatsten weer opgebouwd uit een of enkele groepen. Elk van deze eenheden is gekarakteriseerd door dichtheid, grootte en de mate waarin het verloop van het broedproces bij de verschillende paren synchroon plaats vindt. Bij de keuze van een nestplaats spelen de structuur van de bodem, de vegetatie, de hoogte t.o.v. de zeespiegel en de aanwezigheid van visdieven, noordse sterns en kokmeeuwen waarschijnlijk alle een belangrijke rol. Grote sterns leven vrijwel uitsluitend van haring, sprot, smelt en zandspiering en zijn dus voedselspecialisten. Hun vangtechniek is erg gevoelig voor slechte weersomstandigheden.

II. Functionele aspecten van het verspreidingspatroon van de nesten.

Om de relatie tussen het verspreidingspatroon van de nesten en het voortplantingssucces nader te onderzoeken is een aantal nesten individueel gemerkt en regelmatig gecontroleerd. Uitkomende kuikens werden voorzien van een ring. Voor elk nest werd het voortplantingssucces uitgedrukt in de vorm van het percentage uitkomende eieren en/of uitvliegende jongen. De nesten werden met betrekking tot de volgende factoren op hun voortplantingssucces vergeleken: (1) legselgrootte, (2) de mate van synchronisatie van eileggen en broeden binnen een groep en (3) binnen de eilandpopulatie, (4) de ligging van de nesten binnen de groep, (5) groepsgrootte, (6) koloniegrootte. De invloed van leeftijdsverschillen tussen de vogels en seizoensinvloeden werden bij de analyse zoveel mogelijk uitgesloten. Met uitzondering van groepsgrootte bleken alle genoemde factoren op een bepaalde wijze positief gecorreleerd te zijn met succesvol broeden. De populatie van de grote stern op Griend bleek te zijn opgebouwd uit twee categorieën broedvogels: Individuen die al eens eerder gebroed hadden ('recidivists') en individuen die voor de eerste maal broedden ('recruits'). De eerste categorie broedde eerder dan de tweede en bracht veel meer jongen groot. Dit verschil werd veroorzaakt door een groter aantal eieren per legsel, door een sterkere synchronisatie van het broeden, door een hogere nestdichtheid en door het broeden in relatief grote kolonies. Ook seizoensinvloeden bleken een rol te spelen.

Gegevens die tussen 1966 en 1972 werden verzameld, toonden aan dat eieren en kuikens verloren gingen door roofvijanden (eieren 14.3% en kuikens 11.7%), hoge waterstanden (eieren 3.8%), onvruchtbaarheid (eieren 6.9%), voedselschaarste, ziekte en lichaamsgebreken (tezamen 33.0% van de kuikens).

Om na te gaan hoe grote sterns zich tegen roofvijanden verdedigen en welke rol het verspreidingspatroon van de nesten daarbij speelt, is uitgebreid onderzoek verricht naar het voorkomen van roofvijanden, de wijze waarop zij hun prooi bemachtigen, de schade die zij aanrichten, en de wijze waarop de grote sterns zich tegen roof verdedigen. Tevens is aandacht besteed aan de mogelijke voor- en nadelen van het broeden in de omgeving van kokmeeuwen.

Als potentiële roofvijanden van de grote stern zijn op Griend de volgende soorten waargenomen: kokmeeuw, zilvermeeuw, grote mantelmeeuw, scholekster, kleine jager, velduil en diverse soorten roofvogels. Alleen de kokmeeuw, de zilvermeeuw en de grote mantelmeeuw richtten belangrijke schade aan in de sternkolonies. Voor deze soorten gold dat slechts een klein aantal individuen zich op roven in de sternkolonies had toegelegd. Het aantal rovende kokmeeuwen nam duidelijk toe in de loop van het broedseizoen. De aantallen rovende zilver- en mantelmeeuwen werden vooral beperkt doordat de rovers elkaar onderling uit de omgeving van de kolonie verjoegen. Hoewel de kokmeeuwen een groot aantal eieren en kuikens roofden, had deze roof vooral betrekking op eieren en kuikens waarvan wij uit andere waarnemingen weten dat zij geringe levenskansen hadden. Het vrij geringe aantal kuikens dat door zilver- en mantelmeeuwen geroofd werd behoorde tot een categorie waarvan de levenskansen wel groot waren. Rekening houdend met deze verschillen, bleek dat de kokmeeuwen ongeveer eenzelfde negatieve invloed hadden op het voortplantingssucces van de grote stern als de zilver- en mantelmeeuwen tezamen. Met betrekking tot het voorkomen van roofvijanden en de hierdoor optredende schade kan de situatie op Griend beschouwd worden als representatief voor de meeste andere kolonies van de grote stern in Nederland.

Grote sterns reageren verschillend op verschillende soorten roofvijanden. Indien uitsluitend de eieren en jongen bedreigd worden, blijven de sterns op het nest zitten en dreigen vanuit die positie. Wanneer echter tevens de volwassen vogels bedreigd worden, vliegen zij op. Zij tonen aanvals- of vluchtgedrag. De frequentie en de intensiteit waarmee aanvallen worden uitgevoerd, is meestal opvallend laag, hoewel ze duidelijk toenemen nadat de vogels met hun jongen de kolonie hebben verlaten.

Visdiefjes en kokmeeuwen reageren op vrijwel alle soorten roofvijanden door van het nest op te vliegen. Roofvijanden die eieren en jongen bedreigen, worden fel aangevallen.

De grote stern verdedigt zijn broedsel op effectieve wijze tegen kokmeeuwen, zilvermeeuwen, mantelmeeuwen en scholeksters. Tegen roofvijanden die tevens een bedreiging vormen voor volwassen vogels worden eieren en jongen minder goed beschermd.

De wijze van reactie van roofvijanden op aanvallen door grote stern, visdief en kokmeeuw hangt vooral af van de frequentie en de intensiteit van die aanvallen. Door de gecombineerde aanvallen van de verschillende soorten wordt een groot aantal soorten rovers met succes uit het broedgebied verdreven. Visdieven en kokmeeuwen beschermen de grote sterns tegen rovende, kok-, zilver- en mantelmeeuwen.

De discussie van deel II gaat over (a) de wijze waarop het verspreidingspatroon van de nesten binnen een kolonie schade door roofvijanden kan verminderen, (b) de voor- en nadelen van het broeden in de directe omgeving van een andere soort en (c) de betekenis die het verspreidingspatroon van de nesten kan hebben als beschermingsmiddel tegen diverse omgevingsfactoren.

Op Griend is het voordeel van samenbroeden met kokmeeuwen vrijwel zeker groter dan het nadeel. In kolonies waar anders dan op Griend ook landroofdieren voorkomen, bieden kokmeeuwen nog meer voordelen. Dit leidt tot de conclusie dat het nestelen in de omgeving van visdieven en kokmeeuwen in het algemeen voor grote sterns voordelig is.

De veronderstelling wordt geuit dat het verspreidingspatroon van de nesten in een kolonie van de grote stern en daarmee verband houdende gedragsaspecten geëvolueerd zijn onder de druk van tenminste drie verschillende factoren, nl. voedsel, predatie en getijdenbewegingen. Men zou zich kunnen voorstellen, dat voordelen die samenhangen met gemeenschappelijk voedselzoeken het nestelen in kolonies hebben bevorderd. Een kolonie broedende vogels trekt echter gemakkelijk roofvijanden aan en dat kan geleid hebben tot het kiezen van een broedgebied waarin landroofdieren schaars zijn. Het door de grote stern gekozen gebied introduceerde weer het gevaar dat eieren en jongen door springvloed worden overspoeld. Dit kan op zijn beurt weer van invloed zijn geweest op het ontstaan van speciale gedragsaanpassingen zoals, weinig of geen binding met de broedplaats van het vorige jaar, het kiezen van een relatief hooggelegen nestplaats en op de mogelijkheid van de kuikens om binnen het broedgebied grote afstanden af te leggen. De hoge nestdichtheid in een kolonie grote sterns zou geëvolueerd kunnen zijn onder invloed van de noodzaak om synchroon te broeden. Een tekort aan nestplaatsen die voldoende veiligheid bieden bij hoge waterstanden kan echter ook een positieve invloed hebben uitgeoefend op het ontstaan van een hoge nestdichtheid.

III. Causale aspecten van het verspreidingspatroon der nesten.

In de periode die voorafgaat aan het leggen van de eieren maken de grote sterns op Griend drie opeenvolgende stadia door die gekarakteriseerd zijn door een toenemende groepsdichtheid en een toename van nestel- en agressieve handelingen. Individuen die een nestplaats zoeken vestigen zich bij voorkeur zo dicht mogelijk bij broedende soortgenoten. Het optreden van agressief gedrag zorgt er echter voor dat naburige paren elkaar niet al te dicht kunnen naderen. Broedende vogels dreigen veelvuldig tegen nestplaatszoekende individuen. Het omgekeerde komt weinig voor; de zittende houding van een broedende vogel heeft waarschijnlijk een remmende werking op het agressieve gedrag van de nestplaatszoekers. Verschillen in de afstanden tussen de nesten van synchroon en asynchroon broedende vogels worden verklaard uit gedragsverschillen tussen buurparen op het moment van vestiging. Experimenten met polyesther kunststerns hebben aangetoond dat de grootte en de dichtheid van een groep grote sterns van invloed is op de aantrekkingskracht die zo'n groep op soortgenoten uitoefent. De aard van de reacties waarmee de sterns op de kunstvogels reageren is nader onderzocht.

In een discussie wordt ingegaan op de mogelijke factoren die een rol kunnen spelen bij het tot stand komen van synchroon broeden binnen een groep en binnen de eilandpopulatie als geheel.

Er zijn sterke aanwijzingen dat grote sterns bij het kiezen van een nestplaats worden aangetrokken door de visdieven en de kokmeeuwen. Een poging om de verspreiding van de sterns op Griend te beïnvloeden door polyester kunst-kokmeeuwen neer te zetten bleef zonder succes. Het respectievelijk negatieve en positieve resultaat dat met de op gelijke wijze vervaardigde kunstmeeuwen en kunststerns werd behaald heeft geleid tot de veronderstelling dat grote sterns in de onderzochte situatie aan kunstmeeuwen hogere eisen stellen wat gelijkenis met de levende vogel betreft dan aan kunststerns. De mogelijke oorzaken hiervan worden besproken.